Proceedings of the Institution of Mechanical Engineers

International Conference

Part-load Pumping Operation, Control and Behaviour

1–2 September 1988
Heriot-Watt University
Edinburgh

Sponsored by
Power Industries Division of the Institution of Mechanical Engineers

Co-sponsored by
Japan Society of Mechanical Engineers
Verband Deutscher Maschinen – und Anlagenbau
British Pump Manufacturers Association

IMechE Conference 1988–5

MEP

Published for the Institution of Mechanical Engineers by
Mechanical Engineering Publications Limited

Contents

The publications of

THE INSTITUTION OF
MECHANICAL ENGINEERS

are regarded as core material in engineering
libraries throughout the world.

For details of our

Books

Conference Proceedings

Journals

send for a catalogue to:
**Sales Department,
Mechanical Engineering Publications Limited,
PO Box 24, Northgate Avenue, Bury St. Edmunds,
Suffolk, IP32 6BW England.
Tel (0284) 763277 Telex: 817376**

Publishers to
The Institution of
Mechanical Engineers

The Institution of Mechanical Engineers

The primary purpose of the 76,000-member Institution of Mechanical Engineers, formed in 1847, has always been and remains the promotion of standards of excellence in British mechanical engineering and a high level of professional development, competence and conduct among aspiring and practising members. Membership of IMechE is highly regarded by employers, both within the UK and overseas, who recognise that its carefully monitored academic training and responsibility standards are second to none. Indeed they offer incontrovertible evidence of a sound formation and continuing development in career progression.

In pursuit of its aim of attracting suitably qualified youngsters into the profession — in adequate numbers to meet the country's future needs — and of assisting established Chartered Mechanical Engineers to update their knowledge of technological developments — in areas such as CADCAM, robotics and FMS, for example — the IMechE offers a comprehensive range of services and activities. Among these, to name but a few, are symposia, courses, conferences, lectures, competitions, surveys, publications, awards and prizes. A Library containing 150,000 books and periodicals and an Information Service which uses a computer terminal linked to databases in Europe and the USA are among the facilities provided by the Institution.

If you wish to know more about the membership requirements or about the Institution's activities listed above — or have a friend or relative who might be interested — telephone or write to IMechE in the first instance and ask for a copy of our colour 'at a glance' leaflet. This provides fuller details and the contact points — both at the London HQ and IMechE's Bury St Edmunds office — for various aspects of the organisation's operation. Specifically it contains a tear-off slip through which more information on any of the membership grades (Student, Graduate, Associate Member, Member and Fellow) may be obtained.

Corporate members of the Institution are able to use the coveted letters 'CEng, MIMechE' or 'CEng, FIMechE' after their name, designations instantly recognised by, and highly acceptable to, employers in the field of engineering. There is no way other than by membership through which they can be obtained!

C330/88

Pump instabilities at partial flow—a review

A ENGEDA, PhD, MASME and M RAUTENBERG, PhD, MASME
Institute of Turbomachinery, University of Hannover, Hannover, West Germany

SYNOPSIS   This paper reviews the physical flow conditions at partial flow, which consist of flow separation, back flow, prerotation and rotating stall; and the various attempts that have been made, to explain the occurrence of various forms of partial flow instabilities using these flow conditions.

1. INTRODUCTION

Centrifugal pump design practices are generally empirical in nature, i.e. scaling from existing good designs or from charts and tables. These methods are capable of producing pumps of excellent performance, but they have the disadvantage, that they do not provide clues as to what went wrong with a given design, especially concerning problems of partial flow. In view of current trends to design pumps of variable speed, higher speed, increased pump size, power, head per stage and range of operation, serious instability problems have arisen, especially at partial flow. These problems are very common in pumps for hydro and thermal power plants.

A review of recent literature shows that at least three indirectly related phenomena are involved at partial flow: appearance of reverse flow at impeller inlet and outlet (recirculation), the flow in the suction pipe starts to prerotate with the impeller (prerotation), and flow separation. Even though the effects of this mechanism have been observed for many years, proper explanations for it are few and the effects are not fully understood. Detailed experimental investigations by many workers show that the flow field to be extremely complex, involving non axially symmetric flows and time dependent variables. Accurate prediction of these flow mechanism would require three dimensional flow pattern calculations which are not feasible at this time.

This flow mechanism of recirculation, prerotation, and flow separation imposed on the specific machine geometry and due to the specific interaction of hydraulic parts (inlet to impeller, impeller to recuperator, and pump to piping system), cummulatively or selectively, has the effect of causing the instabilities of partial flow, commonly known as 'droop' curves, radial thrusts, cavitation, surging, and system vibration.

2. PARTIAL FLOW PHYSICS

As flow rate is reduced, the incidence of the flow at the impeller tip, especially at the outermost radius, increases, resulting in a high relative velocity and subsequent rapid pressure recovery on the blade suction side. Figure 1 shows the effect of flow variation above and below the design flow, at the impeller inlet. The effect of continuous flow reduction is higher incidence causing high blade loading on the suction side, and eventually reaching the blade stalling limit which is followed by local separation.

Flow separation in the impeller is generally recognized as a limiting factor. At reduced flow the relative inlet velocity approaches the impeller tangential velocity while at the outlet the relative velocity decreases to zero, which means at some point the critical diffusion ratio is reached. Impeller internal flow solutions require an extensive geometric description of the blading but often impeller design is made using reduced numbers of variables, commonly: specific speed, suction specific speed, diffusion limitations and exit flow angles.

Most previous works suggest that at partial flow, separation, backflow, and prerotation take place in sequence. Observations date back to 1900. Spanhake (1) and Pfleiderer (2), attributed inlet recirculation to non uniform, spanwise total pressure distribution. Probably the first early attempt to explain and model back flow (recirculation) and prerotation is that of Stepanoff (3), where he stated that the cause of inlet back flow is related to prerotation, on the basis of his theory of the principle of least resistance. Janigro et al. (4) carried out experimental studies on several centrifugal pumps and concluded that the cause of back flow to be boundary layer separation. Schiavello et al. (5) hold the same view, and state that the aerodynamic pressure field, meridional flow diffusion and secondary flows could be assumed to be prevailing factors on flow separation and back flow onset. An interesting experimental result is that of Tanaka (6) who investigated a series of impellers to establish the interrelation between inlet and outlet back flow, where he concluded that

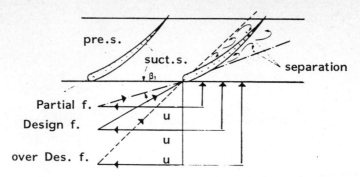

Fig 1   Incidence variation and separation

there exists no direct interdependency. Breugel-mans et al. (7) carried out experimental studies to determine the effect of design parameters on back flow, and reached the same conclusion as Janigro et al. (4), that impeller separation is the triggering mechanism of back flow. The most practical contribution is that of Fraser (8), who attempted to provide calculation procedures to predict back flow. The latest and very much different from previous works is that of Haupt et al. (9), who on the basis of experimental investigation on a centrifugal compressor, argue that inlet back flow is an extension of outlet back flow and infact that this extended back flow is the triggering mechanism for rotating stall. One of the earliest attempts to explain the mechanisms of partial flow through a flow model is that of Polikoviski et al. (10), which is shown in figure 2.

2.1 Flow Separation at Partial Flow

In an impeller passage the maximum velocity is not in the centre of the passage, but nearer to the wall with greater convex curvature. But this is so only if flow separation does not occur, otherwise the velocity maximum can be nearer the other wall, which may have smaller convex or even a concave curvature. As many investigators have shown, on the wall with flow separations, reverse flow may set in. This situation can prevail in some impellers, even at design flow. It is well established that when flow separation occurs, the coriolis forces can set flows into motion which go crosswise to the main flow.

Even though the triggering mechanism of impeller flow separation at partial flow has been the subject of many studies for many years, both in pump and compressor technology, and the necessity for reliable information is ever increasing, there is still no clear established knowledge.

It is logical to search for the design geometric and fluid dynamic parameters that affect separation, in an attempt to reduce and control the instabilities of partial flow ('droop' curves, radial thrusts, various form of cavitation, surging, and system vibration). However, there is again an implication that when boundary layer separation is directly or indirectly the cause of the instabilities, then the relative velocity variation is the key factor.

2.2 Back flow (internal recirculation)

Back flow or recirculation is defined as flow reversal at either the inlet or the outlet tips of the impeller vanes. Various authors have reported on inlet back flow, considering it as a locally generated phenomena. Figure 3 shows a typical representation of inlet back flow as a local phenomena.

Others have reported on the existance of back flow at the impeller discharge and hinting its source to be a local mechanism. Beginning from the results of Peck (11), even though his case was that of at and near shut-off, there is a new trend, also hinted by Fraser (8) for a certain inlet to outlet diameter ratio, that inlet and outlet back flow may be one continuous back flow. This trend of

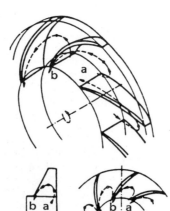

- flow separation at 'a', pressure side

- spiral back flow in channel

- back flow leaves impeller inlet at 'b'

Fig 2   Polikoviskii's partial flow model

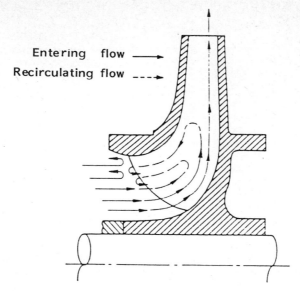

Entering flow ⟶
Recirculating flow ⤍

Fig 3   Inlet back-flow, Karassik's representation

back flow behaviour is the latest argument pro-
moted by Haupt et al. (9) on the basis of investigation on a centrifugal compressor. Figure 4
shows inlet and outlet back flow as independent
flows, Tanaka's (6) representation; and figure 5
is Peck's (11) representation of continuous back
flow.

2.3 Prerotation

Associated with the beginning of back flow at
reduced flows, is the phenomena generally known
as prerotation at inlet. Prerotation is characterized by non axi-symmetric pressure distribution
around the periphery, a negative axial (reverse)
flow and tangential velocity components at the

outer radius, and forward flow in the inner pipe.
Figure 6 shows Breugelmann et al's (7) representation of inlet back flow and prerotation.

The general consensus in the past, seen in the
literature, is that prerotation is induced by the
back flow at inlet. Prerotation is 'positive' when
the fluid rotatin is in the direction of the impeller rotation. It is known that certain inlet configurations (unstable flows in bellmouth or vortices induced by sharp bends) cause negative prerotation, which suggests prerotation could be
basically a result of symmetry loss.

The works of Schiavello et al. (5), Breugel-
manns et al. (7) and Fraser (8) are at the moment
the best works in the literature that attempt to
analyse the causes of back flow and prerotation, and
their likely influence on partial flow instabilities.

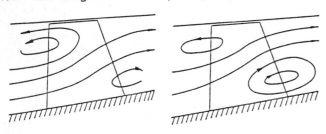

Fig 4   Tanaka's representation of back-flow

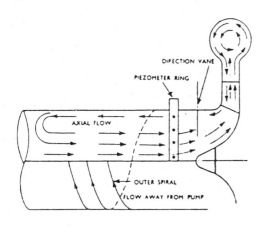

Fig 5   Peck's representation of continuous back-flow

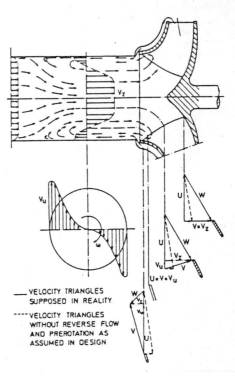

— VELOCITY TRIANGLES
SUPPOSED IN REALITY.

---- VELOCITY TRIANGLES
WITHOUT REVERSE FLOW
AND PREROTATION AS
ASSUMED IN DESIGN

Fig 6   Breugelmann's representation of back-flow and
prerotation

2.4 Rotating Stall

Classically stall is explained on the basis of inlet stalling. If the pump inlet blading is treated as a cascade row, and is tested over a range of incidence (reduced flow), stalling occurs as the incidence is increased. Due to the non uniformity of either the individual blade passages or the entering flow, one passage stalls before the others, in this case passage B in figure 7, taken from Ferguson (12). Flow breakdown in passage B causes a deflection of the inlet fluid stream so that passage C receives fluid at a smaller angle of incidence and passage A at a larger incidence. Then passage A stalls, resulting in a reduction of incidence onto passage B, which then comes out of stall, thus stall rotates around the blade row. Stall manifests itself as a low frequency pressure pulsation, usually at a frequency 2/3 the shaft frequency and is considered to be a constant frequency problem.

Recent deeper knowledge of rotating stall, Casey (13), attributes the rotating stall inception to: incidence stall due to separation on leading edge, blade stall due to separation of the blade profile boundary layers, and wall stall due to separation of the casing or hub boundary layers. Although a great deal of research has been conducted into rotating stall, especially on axial fans and compressors, it is not possible to predict its occurrence with any accuracy.

There have been many attemps to trace some of the partial flow instabilities (Q-H curve instability, inlet surging, back flow, etc.) to rotating stall, but as yet there is no conclusive evidence. Based on experimental work and literature survey, Janigro et al. (4) concluded that rotating stall patterns could be found as a starting point of the back flow. Haupt et al. (9) are certain in their claim that the back flow, which occurs from the outlet of the impeller along the shroud surface (a centrifugal compressor), plays the central role for initiating the inception of rotating stall.

3. FLOW INDUCED INSTABILITIES

It is often reported that at partial flow, especially below recirculation flow (back flow), a number of unfavourable pump conditions arise, which are simply termed partial flow instabilities.

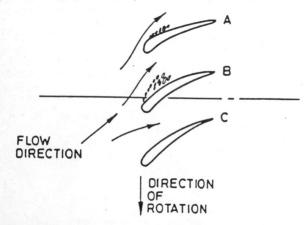

FLOW DIRECTION

DIRECTION OF ROTATION

Fig 7   The mechanism of rotating stall [ref (12)]

The severity of the problem seems to depend on many factors: machine size, design features, material of construction, and the length of this critical operation.

The general trend of analyses of the problem of partial flow instabilities so far has been to link the problem to the start of back flow (recirculation) and prerotation, and as a result of that, all attempts had concentrated on the understanding and controlling of those flow mechanisms.

3.1 "Drooping" Curve

Each pump has a characteristic head-capacity (H-Q) curve which depends mainly on the geometrical dimensions and speed of the machine, i.e. the specific speed. "Drooping" curve or unstable characteristic at partial flow is an H-Q curve that falls to shut-off, i.e. negative slope. An adverse drooping causes system surge, which is common in centrifugal compressor practices, however, such pumps are rarely designed today. Some degree of drooping is present in most centrifugal pumps, of lower specific speed. Yedidiah (14) describes small degree of drooping as harmless and advices against attempts to eliminate it.

Schweiger (15) attributes drooping to prerotation and back flow (recirculation), on grounds that these flow mechanisms are the source of loss that reduce the head. He concludes that design changes which reduce back flow and prerotation will result in reduced drooping. On the contrary Yedidiah (16) reports that this is not always true and claims that on the basis of his experience, an increase of back flow (recirculation) could also result in reduced drooping in certain cases.

Many visual investigations of the inlet pipe flow at shut-off have shown, that a strong vorticity could be generated, which implies, if not fully then partly, the head loss at partial flow is due to back flow.

3.2 Radial Thrusts

Flow conditions in recuperators (volute, vaned or vaneless diffuser) of centrifugal pumps have been investigated both analytically and experimentally for many years. In particular, it is well established that, at partial flow operation, static pressure is not constant around the periphery of the impeller, resulting in radial thrusts. The works of Agostinelli et al. (17) and Hergt et al. (18) are probably the best contributions on the topic.

This hydraulically generated radial thrust, as a result of interaction between recuperator device and impeller, could be very damaging in certain cases of partial flow. In usual design practice radial thrust problems are avoided by selecting the right recuperator device. Figure 8 shows the effect of various types of recuperating device on radial thrust, from (19).

Steady and unsteady radial thrusts are known to occur in centrifugal pumps. The resultant of the steady thrust changes its magnitude with flow-rate. The unsteady radial thrusts, sometimes also rotating, may be superimposed on the steady thrust. Rotating thrusts are common in pumps with vaned diffusers, at partial flow.

© IMechE 1988 C330/88

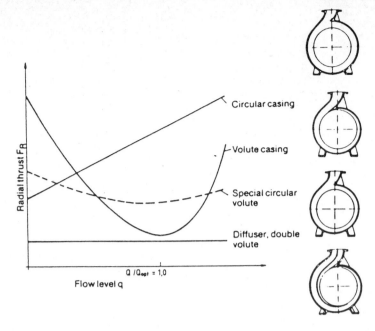

Fig 8   The effect of various casings in inducing radial thrusts [ref (19)]

3.3 "Partial Flow" Cavitation

The increase of head-per-stage, i.e. increases of flow velocities within the pump, and the demand for unrestricted flow range of operation, as low as 25 % of design flow, has resulted among other things in new cavitation problems. Inspite of adequate NPSH some pumps have shown cavitation damage on the pressure side at the inlet and outlet of the vane tip, at the transition section between vane/shroud and vane/hub, and on the suction side of the inlet vane tip, all at partial flow operation.

Fraser (8) reported with certainty that the cavitation damage at outlet and inlet tip on the pressure side can be traced directly to the operation of the pump in the recirculation zone. It is argued that the back flow causes the formation of vortices with high velocity zones at their core, which results in a drop in local static pressure and if it drops below the critical value, it leads to cavitation.

In the impeller channel, if the vane/shroud and vane/hub transition section have large radii, flow direction deviation could be caused resulting in separation which in turn results in cavitation. Figure 9 shows such form of cavitation, from Florjancic(20).

In pumps with over designed inlets (high-suction-specific-speed), sometimes refered to as pumps of the 1950's and 1960's, a cavitation due to recirculation is reported to take place at the impeller inlet tip. Samarasekera (21) claims this cavitation is on the suction surfaces, while Florjancic (20) and Karassik (22) state that it is on the pressure side.

Cavitation due to various causes do occur at partial flow, some of them very destructive. But it seems, despite of many attempts to relate them to recirculation, the mechanisms are not yet fully understood.

3.4 Inlet Surging

In the last thirty years, there has been a strong pressure on the pump designer to reduce NPSH required values, which the designer had to achieve through increases in the impeller suction eye area. According to Karassik (22) this has the effect of bringing the onset of recirculation closer and closer to the best efficiency point, i.e. increasing the instability range at partial flow. Within the recirculation range, there exists a form of instability usually refered to as inlet surging or cavitation surge. This surge is associated with low flow and reduced NPSH values, which results in a large spinning cavity growing and collapsing in the suction pipe, this in turn causes a surging of very high amplitude and low frequency pressure pulsations.

Due to inlet surging when coupled with "soft" systems, boilers or gas-filled surge tanks, a very destructive oscillation could be generated. Since prerotation and back flow are present during inlet surging, there have been many attempts to establish a relationship. There have also been attempts at vibration modelling surging, i.e. as a vibrating mass, a spring-like component, and means by which energy is fed to the vibration. As with rotating stall, the mechanism of surge is not fully understood.

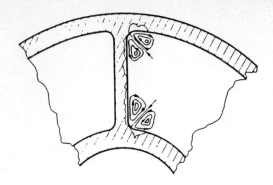

Fig 9   The mechanisms of corner cavitation [ref (20)]

4.   CONCLUSIONS

A review of literature shows that, concerning the physical flow conditions of partial flow which are characterized by flow separation, back flow (inlet and outlet recirculation), prerotation and rotating stall; there is much contradiction and confusion. Therefore it is not surprising that there has been much difficulty in identifying and separating the causes from the effects.

The difficulty in controlling the so called partial flow instabilities: drooping curves, radial thrusts, various forms of cavitation and inlet surging, is due to absence of proper knowledge about the flow conditions mentioned above.

REFERENCES

(1)   Spanhake, W., Centrifugal Pumps, Turbines and Propellers, MIT Press, 1934.

(2)   Pfleiderer, C., Die Kreiselpumpen, Springer Verlag, 1961.

(3)   Stepanoff, A.J., Pumps and Blowers: Two Phase Flow, John Wiley and Sons, 1965.

(4)   Janigro, A. and Schiavello, B., Prerotation in Centrifugal Pumps, Von Karman Institute, LS 1978-3.

(5)   Schiavello, B. and Sen, M., On the Prediction of Preve Flow onset at the Centrifugal Pump Inlet, ASME 25th Int. Gas Turbine Conference, 1980.

(6)   Tanaka, T., An Experimental Study of Back Flow Phenomena in a High Specific Speed Turbomachinery, 10th Int. Conference of BPMA, Cambridge, 1987.

(7)   Breugelmans, F.A.E. and Sen, M., Prerotation and Fluid Precirculation in the Suction Pipe of Centrifugal Pumps, Proc. of 11th Turbomachinery Sympo., Texas A and M, 1982.

(8)   Fraser, W.H., Recirculation in Centrifugal Pumps, ASME Winter Annual Meeting, Washington, Nov. 1981.

(9)   Haupt, U., Chen, Y.N. and Rautenberg, M., On the Nature of Rotating Stall in Centrifugal Compressors with Vaned Diffuser, Part I and II, 1987 Tokyo International Gas Turbine Congress.

(10)   Polikoviskii, V.I. and Levin, A.A., Performance of Pumps and Air Blowers at Reduced Discharge Conditions, Teploenergetika, 1965, 12(1=, 71-74.

(11)   Peck, J.F., Investigations Concerning Flow Conditions in Centrifugal Pumps, Proc. Inst. of Mech. Engrs., Vol. 164, 1951.

(12)   Ferguson, T.B., The Centrifugal Compressor Stage, Buttenworth and Co. Ltd., 1963.

(13)   Casey, M.V., A Mean-Line Prediction Method for Estimating the Performance Characteristic of an Axial Compressor Stage, Int. Conference on Turbomachinery, Cambridge, 1987, Inst. Mech. Engrs.

(14)   Yedidiah, S., Make Pumps With Drooping Curves Your Servants, not Your Enemies, Power, March 1982.

(15)   Schweiger, F., Stability of the Centrifugal Pump Characteristic at Part Capacity, Int. Conference on Pump and Turbine Design and Development, NEL, Glasgow 1976.

(16)   Yedidiah, S., Cause and Effect of Recirculation in Centrifugal Pumps, Part II, World Pumps, October 1985.

(17)   Agostinelli, A., Nobles, D. and Mockridge, C.R., An Experimental Investigation of Radial Thrusts in Centrifugal Pumps, Tor. Eng. Pwr., ASME, April 1960.

(18)   Hergt, P. and Krieger, P., Radial Forces in Centrifugal Pumps with Guide Vanes, Conference on 'Advanced-Class Boiler Feed Pumps', London 1970, Inst. Mech. Engrs.

(19)   KSB, Special print, Centrifugal Pump Design.

(20)   Florjancic, D., Development and Design Requirements For Modern Boiler Feed Pumps, Electric Power Research Institute, Sympo. Proceedings, July 1983.

(21)   Samarasekra, Detection of Inception of Cavitation or Damaging Recirculation in Centrifugal Pumps, World Pumps, December 1984.

(22)   Karassik, I., Flow Recirculation in Centrifugal Pumps World Pumps, 1983-199.

Head and power at closed valve

E W THORNE, CEng, MIMechE
Worthington Simpson, Division of Dresser UK Limited, Newark, Nottinghamshire

SYNOPSIS   The head and power at closed valve are as much dependent on the geometry of the pump as   is the performance at the design point.

Data will be presented and analysed showing how the closed valve performance is controlled primarily by the ratio of impeller outlet area to casing throat area, but with important secondary effects from other aspects of pump geometry.

NOTATION

D	=	Diameter
H	=	Head
Z	=	Number of impeller vanes
b	=	Impeller passage width
U	=	Peripheral velocity
C	=	Absolute velocity
Cu_3	=	Peripheral component outlet velocity
Beta	=	Vane angle
Slip	=	Ideal Cu_3/Actual Cu_3 =

$$1 + (a/Z \times (1 + Beta_2/60) \times (2/(1 - (D_1/D_2)^2)))$$

a	=	Casing factor, taken as 0.77

SUBSCRIPTS

1	=	Impeller Eye diam.
2	=	Impeller Outside diam.
3	=	Impeller outlet
o	=	Closed valve condition
m	=	Meridional component

1   INTRODUCTION

Most analyses of the performance of centrifugal pumps have concentrated almost exclusively on the condition of best efficiency (BEP). However, one of the advantages of centrifugal pumps is their flexibility in being able to operate at part load, and the ability to predict the full pump performance characteristic is desirable.

It is the purpose of this paper to show how the geometry of the pump, as selected to meet a design point, also controls the performance at closed valve.

2   BEP ANALYSIS PROCEDURE.

The analysis procedure for BEP used by the author is that of equating the performance of the impeller to the free vortex behaviour of the casing, as developed by Worster (1), and relating to the empirical deductions of Anderson (2).

Worster used the theoretical closed valve head derived by Busemann (3), to set the slope of his theoretical impeller line, but this approach is only valid for the BEP flow.

The results of the BEP analysis by this method are expressed as:-

1)   A head coefficient/hydraulic efficiency

2)   A flow coefficient

Both coefficients are correlated against the area ratio:-

Impeller outlet area/casing throat area

Thorne (4) later identified a number of minor factors of pump geometry not previously covered.

Since the overall pump geometry has been shown to control the BEP performance, it seemed reasonable to suppose that it would also have a controlling effect on the closed valve performance.

3   CLOSED VALVE HEAD

Closed valve, ie zero flow out of the pump, does not mean that flow within the pump is also zero. Many studies have shown that the closed valve condition is much more complex than at BEP. It appears as though the impeller can only function at one flow rate, and when the casing outlet is closed excess flow out of the impeller is recirculated back through parts of the impeller. It is this recirculation that destroys energy and hence reduces the closed valve head below its theoretical value, and absorbs more power than that due to disc friction and basic mechanical losses.

Pfleiderer(5) gives a complex expression for calculating the closed valve head:-

$$H_0 = (U_2)^2/g \ (1/slip - Clm/2U_1 \ (A^2 + B^2/slip^2))$$

in which A is the ratio inlet diam/outlet diam, and B the outlet diam/cutwater diam.

It is instructive to examine the effect of some of the variables in this equation on the theoretical closed valve head.

In the comparison shown in Table 1 at the end of the paper, the value of the casing coefficient 'a' has been chosen to result in exact agreement between the theoretical and actual closed valve head for the standard geometry(the value 0.77 lies within the usual range), and then various aspects of the geometry changed so that the calculated effect can be seen.

The effect of changing the number of vanes in the impeller has been studied by many authors, including Worster (1) and Varley (6). Their work shows a fall in head coefficient with a fall in the number of vanes, at BEP. Looking at our own data from groups of impeller used in the same casing indicates a similar, but smaller trend at closed valve also, as shown in the table below:-

Table 2   Closed valve head coefficients

PUMP	Ns	AREA RATIO	No of Vanes	HEAD COEFF
A	1500	2.03	6	1.17
			4	1.16
B	850	4.95	6	1.23
			4	1.12
C	1600	2.5	6	1.28
			5	1.22
			4	1.18
D	2700	1.3	6	1.14
			4	1.07

Many more examples could be given but these serve to demonstrate that the effect of a reduction in the number of vanes is similar to that predicted by Pfleiderer's equation above.

Another factor is the use of twisted vanes coupled with the extension of the vane down into the impeller eye. Cutwater clearance is clearly only a minor influence.

A simple head coefficient, $2gH_0/U_2^2$, that is, omitting the hydraulic efficiency term and using the closed valve head, can be plotted against Area Ratio as in Fig.1. to provide a practical example of how a wide range of real pumps behave. This plot shows a clear trend of decreasing coefficient with decreasing area ratio.

The number of vanes has been indicated by symbols in the plot in Fig.1, and helps the effect of these secondary influences to be identified.

Surprisingly, in view of its importance in determining closed valve power, see below, impeller width/impeller diameter has only a small influence on the head coefficient.

The fall in closed valve head coefficient with decreasing Area Ratio is a similar, although less rapid trend, to that at BEP.

Typical ratios of the closed valve head over the BEP head are shown in Table 3:-

Table 3   Variation in head ratio with area ratio

AREA RATIO	Ho/Hn
9	1.1
4	1.14
2	1.16
1	1.25
0.5	1.36

This means that for a given specific speed, the HQ curve gets steeper as Area Ratio decreases, and provides a valuable guide to getting a particular characteristic curve.

Reducing the number of vanes also has a clear effect on the steepening of the HQ curve, since the fall in closed valve head is less than that at BEP, and this offers a further means of fine tuning a charateristic.

4   CLOSED VALVE POWER

A major effect on closed valve power is the b2/D2 ratio as suggested by Mockeridge, and presented in Fig 10.13 of Stepanoff (7). This showed on an abscissa of the b2/D2 ratio a coefficient defined as:-

$$Kcv = \text{closed valve BHP}/N^3/(D_2)^5 \ x \ 10^{-14}$$

A plot of our own data in the same way results in Fig.2. The solid line from Fig. 10.13 is shown for reference.

Most of our data lies below this line, but then so does all of Mockeridges, and his data is contained within our scatter.

Looking at data from the same group of pumps as in Table 2 shows up a similar trend for closed valve power to fall with fewer vanes in the impeller, as shown in Table 4 below:-

Table 4   Variation of Kcv with number of vanes

PUMP	Ns	AREA RATIO	No. of vanes	Kcv
A	1500	2.03	6	1.34
			4	1.28
B	850	4.95	6	0.67
			4	0.47
C	1600	2.5	6	2.02
			5	1.70
			4	1.43
D	2700	1.3	6	2.53
			4	2.02

This is the opposite effect to Stepanoff's explanation to Mockeridges work, in which he suggests that fewer vanes waste more power by allowing more recirculation.

In spite of the trend shown by individual pumps in Table 4, the number of vanes is clearly not a major influence on the closed valve power overall.

Another parameter affecting the closed valve power, is the ratio (inlet whirl/inlet peripheral speed). Thus it generally follows that pumps designed with low inlet velocities, for high suction specific speeds, will have higher closed valve power consumption. Conversely, the acceptable range of operation of pumps of low suction specific speed will be wider as such impellers are inherently more reliable at low flows due to the lower amount of recirculation.

The picture for closed valve power therefore remains imprecise, and for a given design it is necessary to relate to other similar pumps to refine the predictions.

5   COMPLETE CHARACTERISTIC PREDICTION

Use of the closed valve data given above, together with the BEP data given by Anderson and Thorne, enables the overall influence of the pump geometry on the complete characteristic to be determined.

Fig.3 shows four alternative estimated performance curves for a given specific speed, obtained by varying the pump geometry.

Table 1   Closed valve head predictions

D2	203.00	203.00	203.00	203.00	203.00	203.00
D1	64.00	64.00	70.00	70.00	64.00	64.00
RPM	2900.00	2900.00	2900.00	2900.00	2900.00	2900.00
D4	215.00	215.00	215.00	220.00	215.00	215.00
Com	3.32	3.32	2.78	2.78	3.32	3.32
No.vanes	6	5	6	6	6	6
BETA2	27.00	27.00	27.00	27.00	22.00	18.00
a	0.77	0.77	0.77	0.77	0.77	0.77
SLIP	1.41	1.50	1.42	1.42	1.39	1.37
Ho	59.49	56.50	61.00	61.25	60.41	61.16
Ho coeff	1.23	1.17	1.26	1.26	1.25	1.26
Ho test	59.50					

In these pumps, the area ratio has been progressively increased, and the proportions adjusted each time to keep the BEP duty to 100 m^3/h x 50m total head.

The main dimensions of these 4 pumps are shown on the curves. It should be noted how, as area ratio increases, the pump shape changes, the HQ curve gets flatter, the efficiency rises, and the power curve changes from non-overloading to continuously rising.

It can therefore be seen how, combined with knowledge of the operating methods and system, pumps can be closely matched to the system by the methods described above, and potential problems of partload operation avoided.

ACKNOWLEDGEMENT

The author wishes to thank the Directors of Worthington Simpson for permission to publish this paper.

REFERENCES

(1) WORSTER, R. C. The flow in volutes and its effect on centrifugal pump performance. Proc.I.Mech.E. Vol.177, No.31 (1963)

(2) ANDERSON, H. H. Modern developments in the use of large single entry centrifugal pumps. Proc.I.Mech.E. Vol. 169, No.6 (1955)

(3) BUSEMANN, A. The delivery head of radial centrifugal pumps with logarithmic spiral blades. Z. angew. Math. Mech. Vol. 8 No. 5 (1928)

(4) THORNE, E. W. Design by the area ratio method. BPMA 6th Conference, Canterbury (1979)

(5) PFLEIDERER, C. Die Kreiselpumpen. Springer, Berlin. (1961)

(6) VARLEY, F. A. Effects of impeller design and surface roughness on the performance of centrifugal pumps. Proc.I.Mech.E. Vol. 175, (1961)

(7) STEPANOFF, A. J. Centrifugal and axial flow pumps. John Wiley, New York. (1957)

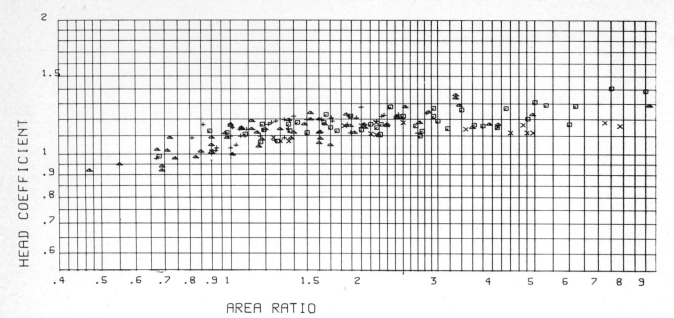

Fig 1   Variation of closed valve head coefficient with area ratio

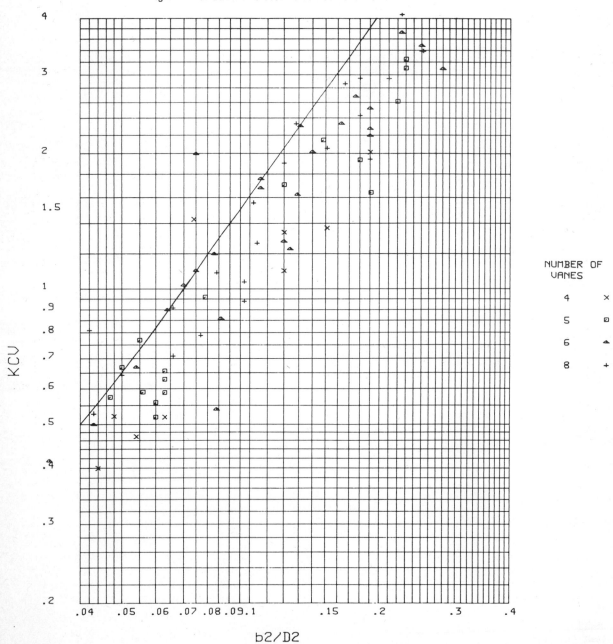

NUMBER OF
VANES

4   ×

5   ▣

6   ▲

8   +

Fig 2   Variation of closed valve power coefficient with ratio b_2/D_2

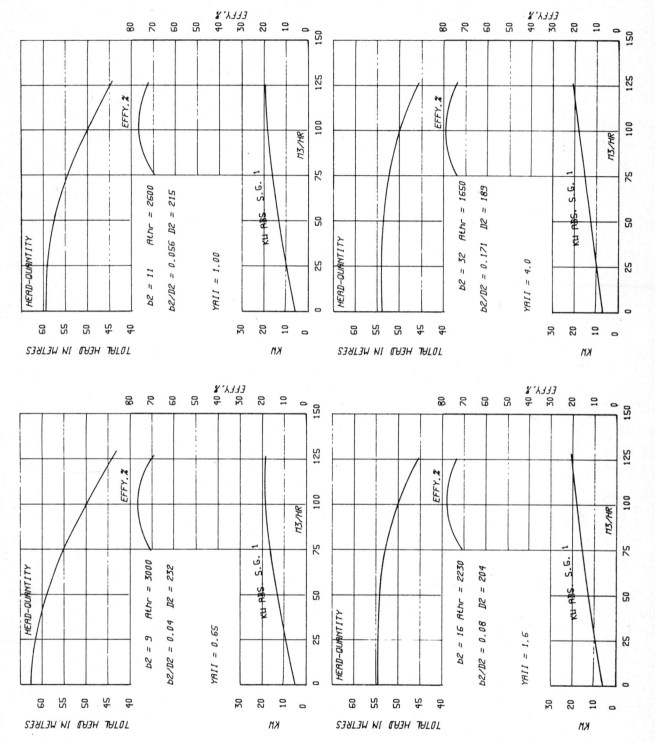

Fig 3   Complete characteristics for four different area ratios

C332/88

Off-design performance of a mixed-flow pump

S SOUNDRANAYAGAM, BSc, PhD
Indian Institute of Science, Bangalore, India
V RAMARAJAN, ME, PhD
Fluid Control Research Institute, Palghat, India

SYNOPSIS   Some results of detailed flow measurements in a mixed flow pump are presented and are discussed to elucidate the physics of the flow at part load conditions.   The most noticeable feature is the occurance of recirculation and its consequences when the flow rate is greatly reduced.   The necessity for making head correction at both inlet and outlet at low discharges is pointed out.

NOMENCLATURE

C	Absolute velocity	(m/s)
C_N	Velocity normal to probe	(m/s)
C_P	Velocity parallel to probe. Positive outward from pump axis	(m/s)
C_Z	Axial velocity	(m/s)
C_θ	Tangential velocity Positive in direction of impeller rotation	(m/s)
D	Impeller tip diameter at outlet	(m)
H	Pump total head	(m)
H'	Normalised head	$[=H/N^2D^2]$
N	Pump Speed	(r/min)
N_s	Specific speed	$[=NQ^{\frac{1}{2}}/H^{\frac{3}{4}}]$
P_T	Total head	(m)
P_s	Static head	(m)
Q	Flow rate	(m³/s)
Q'	Normalised flow rate	$[=Q/ND^3]$
R_e	Reynolds number	$[\frac{UD}{\gamma}]$
U	Tip speed	(m/s)
α	Kinetic energy correction factor at pump inlet and outlet. Also flow angles in Fig 2.	
β	Pitch angle. Positive for positive C_P	
η_i	Internal efficiency	
γ	Kinematic viscosity	(m²/s)
θ	Yaw swirl angle. positive for positive C_θ	

1   INTRODUCTION

Mixed flow pumps are widely used for cooling water circulation duties in thermal power stations.   Large variations of water level in the inlet sumps can occur specially when the cooling water supply is drawn from natural lakes.   The pumps are thus called upon to work over a fairly wide range of operation well removed from the so called design point.   At off design conditions specially at flow rates lower than the optimum, flow reversals and recirculations occur that can result in unsteadiness and vibration leading to mechanical failure and in one instance at least to a substantial rise in water temperature. Detailed measurements of flow over a wide range of flow rates and Reynolds number were made in a model mixed flow pump where the primary motivation was the study of scale effects on efficiency [1].   Parts of the measurements that relate to the internal flows and overall performance at off design conditions are discussed here in an attempt to build an understanding of the physics of the flow at conditions removed from the optimum.

2   THE TEST PUMP AND EXPERIMENTAL PROGRAMME.

The measurements were made in a model pump of specific speed N_s = 118 r/min and inlet diameter 155 mm.   The main dimensions of the pump are shown schematically in Fig 1.   The impeller blades were of aerofoil section obtained by the conical transformation of a plane cascade designed using two dimensional blade data.   The impeller was housed in an existing volute of circular cross section.   The traverse locations are indicated in Fig 1. Details of the hydraulic design of the blades are given in Fig 2.

The experiments were carried out in a closed circuit pump test rig described in [1] which also discusses the measures undertaken to maintain accuracy.   The uncertainty in efficiency measurement was estimated to be ± 1.05 per cent to the usual 95 per cent confidence limits.   The large number of test points obtained establish the mean value of efficiency and its variation to a much higher degree of confidence than that applying to an individual reading. Traverses of velocity, yaw and pitch angles, and total and static pressures were made using calibrated four hole cylindrical probes of compact design [2]. The accuracy of the traverses was checked by comparing the measured three dimensional field with the flow indicated by the calibrated test loop flow nozzle. The agreement in most cases was within 1 percent but fell to about 2 percent at very low flow rates.

All the traverses were carried out under non cavitating conditions ensured by controlling the test rig static pressure. Initial traverses were made at approximately one third blade chord (26 mm) away from the impeller exit along positions 1',2' and 3' of Fig 1. There was difficulty in interpreting these measurements because of the diffusion of the jet mixing in the volute. Further traverses were done closer to the impeller exit (3 mm from impeller outlet edge) at positions 2 and 3 (120° apart) as shown in Fig 1, Traverses were also made in the inlet pipe at a station 2 inlet pipe diameters upstream of the pump inlet flange and in the outlet pipe at a station 2 outlet pipe diameters downstream of the outlet flange.

3 PRESENTATION AND ANALYSIS OF RESULTS

3.1 Quality of flow at inlet and exit from pump.

Usually the correction for velocity head is accounted for in pump head calculations by taking the average velocity over the measured cross section. But that may lead to error if the non uniform velocity distribution specially at pump outlet is not considered. A kinetic energy correction factor α for the velocity head due to non uniform distribution is expressed as

$$\alpha = \frac{1}{A} \int_A (c/\bar{c})^3 \, dA \qquad (1)$$

where $\bar{c}$ is the mean velocity given by

$$\bar{c} = \frac{1}{A} \int_A C \, dA$$

The calculation of α using equation (1) is valid only when all the stream velocities are parallel to the pipe axis. In the present measurements this was not the case specially in the delivery pipe and also in the suction pipe at low flows. Hence equation (1) is modified to

$$\alpha = \int_A c_z c^2 \, dA / \bar{c}_z^3 A \qquad (2)$$

The measurements showed that the flow in the inlet pipe was fully developed with a constant value of α = 1.04 except at flow rates below Q' = 0.002 where the value of α shot up. This increase in α was because of the acceleration of the main stream due to the constriction of the pipe flow by flow reversal associated with inlet autorotation. There were differences in calculated discharge, kinetic energy correction factor etc., in some cases as high as 15 percent among the traverses along three diameters in the pump outlet pipe, whereas the values of Q and α calculated over different diameters at inlet were identical. Though the discharge computed along the three separate diameters at the outlet varied amongst each other the average value agreed well with the known discharge. The kinetic energy correction factor was similarly averaged in the hope that it would give equally representative results. The value of α at outlet varied from 1.4 at high flow rates to 1.07 at reduced flow rates, but suddenly shot up when the flow was lowered much below Q' = 0.004.

Considerable swirl existed in the pump outlet pipe. Fig 3 illustrates the measured swirl angle along one selected diameter in the outlet pipe at various flow rates. The swirl distribution along other diameters at the same station was qualitatively similiar but not identical. The swirl was not axisymmetric and the centre of the swirl did not coincide with the pipe centre line. The magnitude of the swirl increased as the flow rate was reduced with the swirl angle becoming greater than 90 degrees near the centre of the pipe at the lowest flow rate measured of Q' = 0.001. This indicated the existance of a concentrated reverse flow region near the pipe centre at that flow rate. Flow reversal with autorotation was measured in the inlet pipe at flows around Q' = 0.002. The reversed flow was confined to an annular region enveloping a forward flowing core, the direction of autorotation being the same as that of the impeller and caused by the angular momentum of fluid from within the impeller moving upstream in the reversal region and mixing with the incoming flow.

3.2 Correction of head flow characteristics.

Flow conditions at the inlet and outlet traverse stations upstream and downstream of the pump flanges have a marked effect on the measured head flow characteristics. The head-flow characteristics at two sample Reynolds numbers are shown in Fig 4. There is a distinct and repeatable effect of Reynolds number at the higher flow rates, but the effect of Reynolds number is very marked below the value of $Q' = 0.003$ which is a little below the flow rate at which inlet autorotation was first measured. It is difficult to comprehend that these marked differences could be due to any hydrodynamic changes in the impeller itself at low Reynolds numbers. As the measured head is the difference between the total head at outlet from and inlet to the pump, it is possible that the apparent changes in head could be due to wrong indications of head at inlet and outlet.

The code 'Acceptance Tests for Centrifugal, Mixed and Axial Flow Pumps' [3] recognises that at low flow rates autorotation at inlet could reach the measuring station and affect the measured pressure. The centrifugal effects of autorotation would indicate a high pressure at inlet resulting in an apparently lowered total head change across the pump. The code suggests a technique for correcting this. There is also the possibility that the reverse flow associated with inlet autoration would constrict the effective flow area, thus accelerating the main flow and dropping the inlet pressure. Measurements at inlet did confirm that the reverse flow reached well into the inlet traverse station. The measured static pressure at inlet however kept rising as the flow decreased showing that the centrifugal effects of autorotation was the dominant factor causing changes in the pressure. The code suggests that the rise in pressure at low flow rates be ignored and the true inlet pressure taken to be that given by the dotted extrapolation of the linear variation at higher flow rates (Fig 5). The intercept of the dotted extrapolation with the zero flow axis corresponded in every case with the level of fluid in the suction tank. When the measured head-flow characteristics were corrected for inlet head variation along the above lines, the resultant curves shown by the dotted lines in Fig 4 differ quite a bit from one another. The variation of outlet head with flow rate is shown in Fig 6 and seems to be approximately linear at the larger flow rates. This linear variation was extrapolated to zero flow rate as for the inlet flow case and a corresponding correction to the outlet head made. The variation of outlet pressure at low flow rates is no doubt associated with the strong swirl present as seen in Fig 3. When corrections for both inlet and outlet head were made as above, the head-flow characteristics collapsed to the chain dotted curves of Fig 4 which lie very close to one another

suggesting that there are no great changes in the internal flow of the impeller itself causing the discrepancy between the originally measured curves. The apparent variation in these curves at low flows is due to misindication of the inlet and outlet heads. It also suggests that an inlet head correction as suggested by the code is not sufficient and that a correction for the outlet head should also be included. The use of the kinetic energy correction factor in computing the velocity head resulted in these experiments in a maximum correction of approximately 1.35 percent in the total head rise at the largest flow rate.

3.3 Variation of part load efficiency and head with Reynolds number.

The variation of internal efficiency with Reynolds number at different flow rates Q' is shown in Fig 7. A distinct hump is noticed in the curves for flow rates below the optimum Q', but no hump at flows above the optimum Q'. Measurements of impeller loss at part load and its variation with Reynolds number [1] show a smooth variation suggesting that the localised humps in Fig 7 do not have their origin in events occuring in the impeller. However the loss variation at part load with Reynolds number in the volute [1] does show a trough at the same Reynolds number as the peak in Fig 7 for flow rates below the optimum but no trough at the optimum or higher flow rates. This suggests a possible explanation for the hump. Due to the mismatch of incidence at lowered flow rates there could be local separation in the volute possibly at the tongue. At very low Reynolds number the separation would be extensive with large losses. The losses would decrease as increase in Reynolds number decreased the extent of separation. At the peak of the hump the separation bubble would be laterally thin with reattachment causing the passage downstream to flow full. The low skin friction in the separated region with the thin separated bubble with reattachment keep the losses low giving a local peak in efficiency. To the right of the peak the increased Reynolds number reduces the length of the separation bubble giving attached flow after a Reynolds number of about 2×10^5, beyond which the friction continues to drop in the whole pump giving a continuously rising efficiency that gradually flattens out.

A noticeable hump is also encountered in the variation of pump head at part load with Reynolds number (Fig 8). The values of head are uncorrected for inlet and outlet swirl at reduced flow rate and Reynolds number. If this were done, the curves for the two lowest flow rates would also show a hump. Similar shaped curves but with a more pronounced hump have been reported by other investigators [4][5][6]. The hump in these curves lies at a much higher Reynolds number than the local peak of efficiency in Fig 7 confirming that the latter is unconnected with the impeller. The humps in Fig 8 are shallow but their peak shifts to lower Reynolds numbers as the flow rate is decreased. A decrease in flow rate increases the blade angle of incidence. The shift in the peak of the humps in Fig 8 is consonant with the observation that the optimum lift drag ratio for some aerofoils occurs at higher incidence as the Reynolds number is decreased (see Fig 70 p.155 and Fig 77 p.163 of [7]). The clue to the shape of the head variation should lie in the boundary layer changes on the blade with possibly a stationary bubble on the suction surface that affects the drag as well as conceivably increasing the blade circulation.

3.4 Flow traverses upstream and downstream of impeller.

Traverses of velocity, flow direction and pressure upstream and downstream of the impeller yield information about the flow processes at part load. Traverses at impeller inlet showed that the tangential velocity component remained zero except when flow reversal took place at the casing when the flow rate was heavily reduced. Reverse flow at inlet took place when the flow was reduced to about $Q' = 0.004$. This reversed region extended further upstream as Q' was reduced, reaching the traverse station 2 diameters upstream of the inlet flange at the severely reduced flow rate of $Q' = 0.002$. The reversed flow which originates from the casing region of the impeller, rotates in the same direction of the impeller as it moves upstream and turns round and joins the main inlet flow forming a recirculating region. Measurements with pressure differential probes as in this experiment would indicate that the recirculating region was axisymmetric, that is, it extended as an annular region completely enveloping the forward flowing fluid in the core of the inlet pipe. However measurements with rapid response instrumentation such as a hot wire in an aerodynamic model of a pump has shown that the recirculating region is not axisymmetric but broken up into distinct cells that themselves rotate relative to the impeller. This latter phenomenon closely related to rotating stall in compressors can be the root cause of much rough running and vibration at part load operation.

An interesting feature was the greatly increased total and static pressure measured at impeller inlet in the recirculation region. This was due to the flow passing in and out of the entry region of the blade because of recirculation and having blade work done on it several times. A steady state increased total pressure is reached as portions of this high pressure recirculating eddy are torn off by turbulent entrainment and pass downstream. This would result in a region of locally increased total pressure downstream of the blades as well at these reduced flow rates. In a recently investigated industrial pump the same action caused a considerable rise in water temperature at impeller inlet.

Sample results of traverses immediately downstream of the impeller for two flow rates, one at $Q' = 0.0059$ which is just above the optimum flow and the other at $Q' = 0.0038$, are presented in Figs 9 and 10. The traverses refer to measurements at station 2 [Fig 1] and closely ressemble the measurements at station 3. The existence of the secondary flow spirals in the volute is clearly seen in the negative value of C_N at the end of the traverse furthest removed from the hub. The whirl velocity C_θ in the volute would be high at large flow rates while it would be low immediately behind the blades at large flow rates due to the lowered incidence and hence low blade loading. As the flow rate is decreased the value of C_θ in the volute region drops while it rises downstream of the blades due to the increasing blade loading. In Fig 9 for $Q' = 0.0059$ the value of C_θ in the volute region and immediately downstream of the blades is of the same order allowing an interesting phenomenaon to be noticed. At the blade tip a local peak in C_θ will be seen at a radius just lower than the tip and a local trough in C_θ at a radius just above the tip. This is due to the tip vortex being shed from the blade tip. When the flow rate is lowered to $Q' = 0.0038$, Fig 10, the value of C_θ in the volute region is lower than downstream of the blades as anticipated. However the value of C_θ behind the blades has fallen near the hub indicating that we may be approaching stall. The increased C_θ near the tip is almost certainly associated with the transport of rotating fluid from the reversal region at the casing at inlet. The flow traverses at inlet had indicated the beginings of flow recirculation near the casing. Considerable velocity C_P parallel to the probe existed at the lowered flow rate, almost of the same magnitude of the meridional normal velocity C_N and directed outwards towards the tip. At $Q' = 0.0059$ C_P is negative at radii greater than the tip which is consonant with the existance of the secondary flow spirals in the volute. The value of C_P in this region at the lowered Q' of 0.0038 is positive indicating that the secondary flow spirals have been pushed out by this strong cross flow directed radially outwards.

The total pressure P_T is high behind the blades and is fairly uniform in the volute region. The traverse in Fig 9 shows a slight drop in total pressure where the flow in the outer volute region meets the flow leaving the blades. A possible reason for this dip is the location there of the core of the tip vortex shed by the blades, which was also noticed earlier in the C_θ variation. The fact that the static pressure variation also shows a corresponding dip lends support to this view. It was suggested earlier that portions of the inlet reversed flow eddy of high total pressure would be torn off and pass downstream. This can be noticed in the total pressure variation measured downstream at $Q' = 0.0038$ (Fig 10) where there is a steep increase in the total pressure at the tip region.

4 CONCLUDING REMARKS.

The shape of the head-flow characteristics show marked changes at reduced flow rates. These changes are affected both by the reduced capacity and Reynolds number. The differences can be reconciled when corrections are made to allow for swirl both at inlet and outlet from the pump. The recommendations in the codes give an inlet head correction at low flow rates. The present results suggest that an outlet head correction should also be made.

Substantial swirl exists in the outlet pipe, increasing as the flow rate is reduced. A small central core of reversed flow was noticed at the lowest flow rates. It is expected that the quantum of swirl in the exit pipe would be less for pumps of lower specific speed and higher for larger specific speeds.

A small localised hump is seen in the variation of internal efficiency with Reynolds number at part loads. This appears to be connected to changes of the flow in the volute. A slight increase in pump head occurs in a particular Reynolds number range at all flow rates. This seems to be related to changes in the blade boundary layer causing changes in the lift to drag ratio.

Inlet flow reversal at reduced flows results in an increase of total pressure at inlet and outlet at the tip region. The flow traverses show the presence of secondary flow in the volute and the tip vortices shed from the blade tips.

REFERENCES:

[1] Ramarajan V and Soundranayagam S. Scale effects in a mixed flow pump Pts I and II Proc Inst Mech Engrs. 1986 Vol 200. No 43 pp. 173-186.

[2] Soundranayagam S and Ramarajan V. A new four-hole cylindrical probe for 3 dimensional measurements. Review of Scientific Instruments 1980 Vol 51 No.7 p.989.

[3] British Standards Institution. Acceptance tests for centrifugal, Mixed and axial flow pumps. B.S. 5316:1976. Part 2.

[4] Nixon R.A. and Cairney W.D. Scale effects in centrifugal cooling water pumps for thermal power stations. N.E.L. report 505. 1972.

[5] Hutton S.P. and Fay A. Scaling up head flow and power curves for water turbines and pumps. IAHR Symposium Vienna. 1974.

[6] Fay A. Theory for the Reynolds number effect on the head and power of hydraulic machines. Proceedings. 5th conference on Fluid Machinery. Budapest. 1976 pp 261-271.

[7] Abbot I.H. and von Doenhoff A.E. Theory of Wing Sections. 1959 Dover. New York.

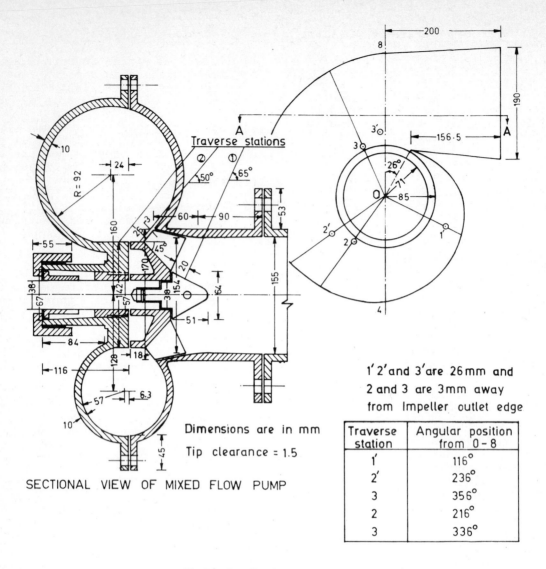

Traverse stations

Dimensions are in mm

Tip clearance = 1.5

SECTIONAL VIEW OF MIXED FLOW PUMP

1' 2' and 3' are 26 mm and
2 and 3 are 3 mm away
from Impeller outlet edge

Traverse station	Angular position from 0–8
1'	116°
2'	236°
3	356°
2	216°
3	336°

Fig 1   Details of test pump

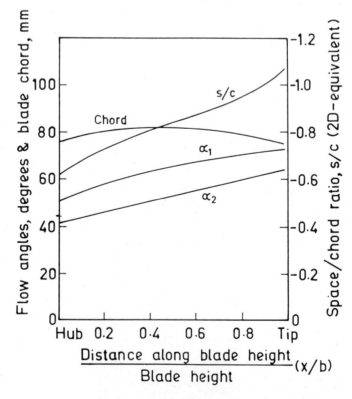

Fig 2   Blade design parameters

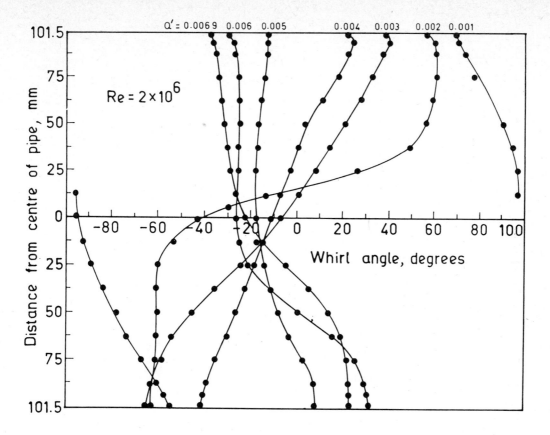

Fig 3   Distribution of swirl angle in outlet pipe

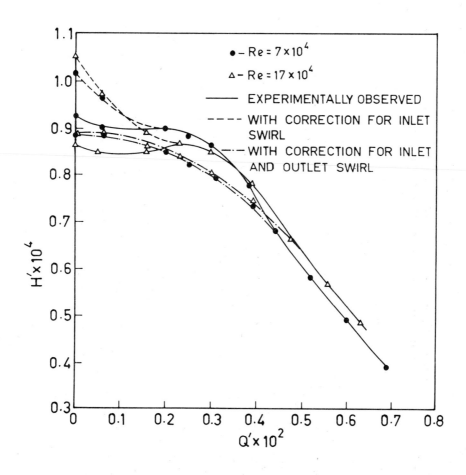

Fig 4   Head flow characteristics at different Reynolds numbers

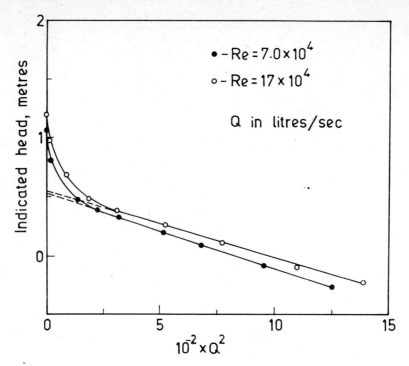

Fig 5   Extrapolation of inlet head at low flowrates

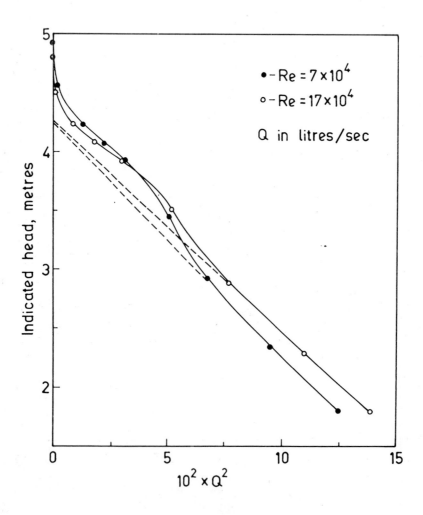

Fig 6   Extrapolation of outlet head at low flowrates

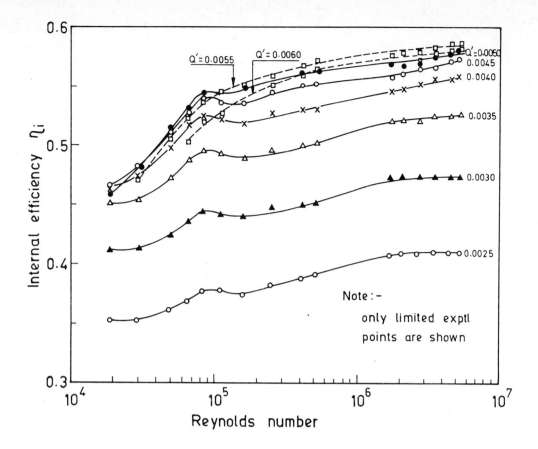

Fig 7   Variation of part-load efficiency with Reynolds numbers

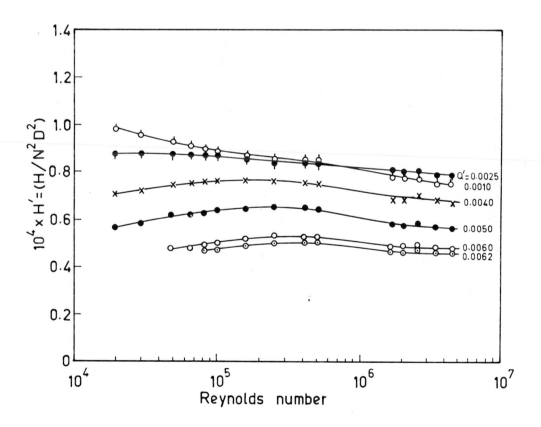

Fig 8   Variation of part-load head with Reynolds numbers

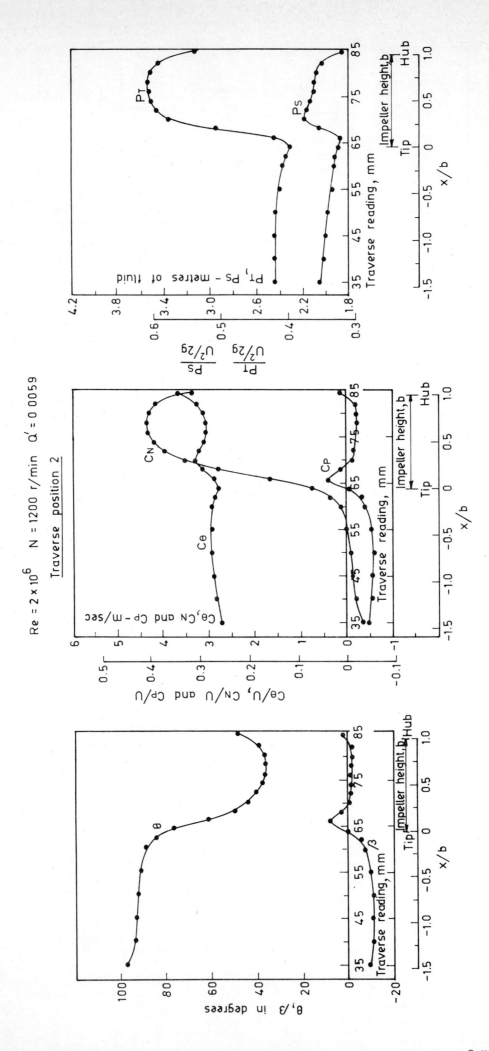

Fig 9   Variation of flow parameters downstream of impeller at $Q' = 0.0059$

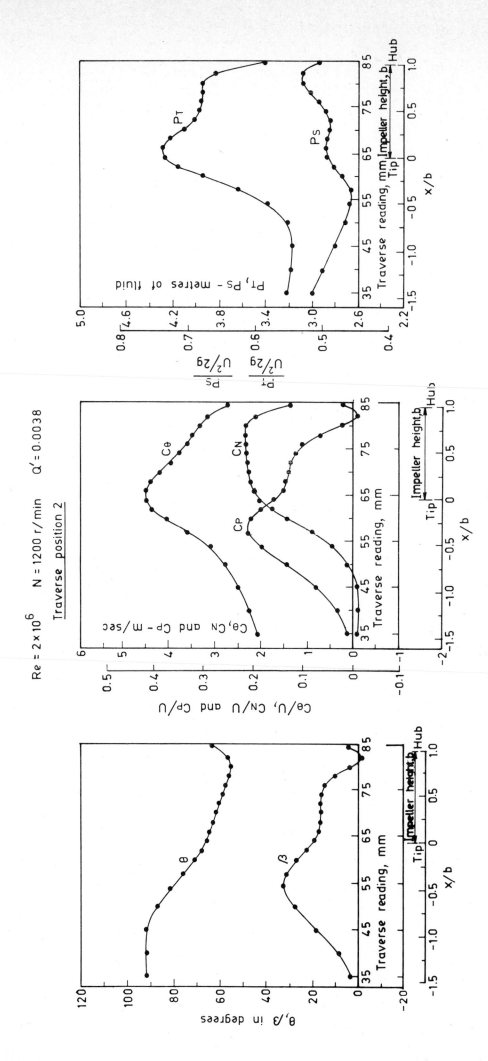

Fig 10   Variation of flow parameters downstream of impeller at Q' = 0.0038

C333/88

Effect of suction duct design on the performance of mixed-flow pump at partial flow

R CANAVELIS, PhD and J-F LAPRAY
Bergeron SA, Paris, France

ABSTRACT

In many modern installations pumps have to operate in off-design conditions. Consequently the manufacturer may have to give guarantees on head and power input at zero flow, as well as on the slope of the head-flow curve at partial flow.

The present study shows how the shape of the head-flow curve may be changed in the case of a mixed flow pump with a specific speed of 4,500 (ns = 90 in metric system, ns = 1.8 in non-dimensional units) by varying the suction duct geometry. 10 different suction lines are tested and a simple method to predict the shape of performance curves is proposed.

1 - INTRODUCTION

The suction duct design of a vertical mixed flow pump is generally determined in order to assure satisfactory suction conditions in the normal operating range : uniform accelerating flow to the bell and absence of vortices on the free surface in the sump.

However in many modern installations, pumps have to operate in off-design conditions and more particularly at partial flow. In such conditions, pump performance curves are influenced by the interaction of inlet recirculation with intake structures.

It is the reason why the manufacturer has to know the influence of suction duct geometry on performance curves shape before giving a guarantee on the following quantities :

- head at zero flow : a limit value is often given by the circuit test pressure.

- power input a zero flow : a limit value is required by the drive characteristics.

- slope of the head-flow curve : a constant negative slope is often required in the contract when several pumps have to operate together.

The present study shows how the shape of the head-flow curve may be changed in the case of a mixed flow pump with a specific speed of 4,500 (ns = 90 in metric system, ns = 1.8 in non-dimensional units) by varying the suction duct geometry. 10 different suction lines are tested and a simple method to predict the shape of performance curve is proposed.

2 - TEST LOOP DESCRIPTION

The main characteristics of the tested pump are :

Nominal flow – Q_N = 7 140 gpm (0.450 m³/s)

Nominal head – H_N = 75.5 Ft (23.0 m)

Rotational speed – n = 1 480 rpm

Impeller inlet diameter - D1= 11.8 in (300 mm)

Suction duct internal diameter - DO = 13.4 in (338 mm)

The pump is installed as shown in figure 1 with different suction ducts or fittings (see figure 2) :

1   – Straight pipe and bellmouth in an axisymetrical chamber

2   – same as 1 with a guide cone

3   – same as 1 with one guiding web (webwidth = $\frac{Ro}{8}$) at a distance L such that L/Do = 0.10 from the impeller (Ro = suction pipe radius)

4   – same as 3 with two opposite guiding webs (Webwidth = $\frac{Ro}{8}$)

5   – same as 1 with two opposite guiding webs (width = $\frac{Ro}{8}$) in the bellmouth at a relative distance L/Do = 2.15 from the impeller

6 - same as 5 with two opposite guiding
    webs (width = $\frac{Ro}{4}$) in the bellmouth

7 - straight pipe with honeycomb straightener
    at a relative distance L/Do = 4.5 from
    the impeller (not shown in figure 2)

8 - same as 1 with two guiding webs upstream
    of the bellmouth at a relative distance
    L/Do = 3.09 from the impeller.

9 - same as 1 with one cross upstream of the
    bellmouth at a relative distance
    L/Do = 2.60 from the impeller.

10 - same as 9 with only one splitter upstream
     of the bellmouth

For each one of these arrangements the complete
performance curves were determined in order to
compare the variations of head H, input power
P and efficiency η , from zero flow to 1.15 Q_N.

The measurement uncertainties are as follows :

head : $\pm$ 0.5 %

Flow : $\pm$ 0.8 %

Torque : $\pm$ 0.2 %

Rotational speed : $\pm$ 0.05 %

3 - TESTS RESULTS

The complete performance curves are given in
figure 3 and 4.
Main results are recalled in table 1 which gives head Ho and input power Po at zero flow as
well as the average slope ΔH/ ΔQ of the Q-H
curve at partial flow. When two values of the
slope are given, one positive and one negative,
it means that the Q-H curve present a maximum.
Table 1 gives also the value of the critical
flow QK at which the recirculating flow reaches
the splitter or the webs installed in the suction duct.

4 - DISCUSSION OF RESULTS

4.1 Variation of curves shape

The comparison of performance curves for the 10
different arrangements shows clearly the development and the influence of inlet recirculating
flow.

For arrangement 1, the "Kink" in the Q-H curve
for a flow of approximately 0,320 m³/s indicates
clearly the occurence of recirculation. The corresponding relative flow rate is Q* = Q/QN =
0.71. The visualization of impeller inlet with
the help of colored threads stuck on the blades
leading edge shows that recirculation appears
indead at this flow-rate.

With arrangement 2 performance curves are very
similar to the previous one. The guiding cone
has nearly no influence on the recirculating
flow.

In arrangement 3 the guiding web with a width
of $\frac{Ro}{8}$ stops partially the recirculating flow
and has a sensitive influence on head and in-
putpower for any flow lower than 0,310 m³/s
(higherH and P than in arrangement 1). The
corresponding relative flow rate is Q* = 0.69.

In arrangement 4 the same tendency is shown
asin arrangement 3. Two webs instead of one
doesnot increase very much H and P at partial
flow.

In arrangement 5 the guiding webs in the bell
mouth are reached by the recirculating flow at
a lower flow-rate of 0.200 m³/s (Q* = 0.44).
For higher flow rates there is no sensitive
difference with arrangement 1. For lower
flow-rate the results are similar with those
obtained in arrangements 3 and 4.

In arrangement 6, it may be noticed that the
effect of wider webs is higher than in case 5,
leading to higher head and power input.

In arrangement 7 it may be seen that the ho-
ney comb straightener located in a straight
suction pipe at a distance L/Do = 4.5 is reached by the recirculating flow at a flow rate
of 0.160 m³/s (Q* = 0.35). For higher flow-
rates there is no sensitive difference with
arrangement 1. For lower flow-rates higher
head and power input are obtained.

Arrangements 8, 9 and 10 show a strong influence of webs, or splitter or cross located
upstream of the bell-mouth for flow-rates
lower than 0.200 1/s (Q* = 0.44).

It is noticable that in all arrangements pump
efficiency curve η is nearly unchanged over
the all operating range from zero flow to nominal point as shown in Figure 4.

Influence of upstream guide vanes on performance curves

Figure 5 gives an idea of flow pattern in the
suction duct at partial flow without any guide
vane (1). Velocities upstream of the impeller
have been measured with a Laser Doppler velocimeter (L.D.V.) as well as with a five holes
probe in sections 1, 2 and 3. (2)

Figures 6 and 7 give some results obtained
with the L.D.V techniques.

At a given partial flow, a reverse flow is
observed in the suction duct near the boundary. This flow corresponds to negative meridional velocities Cm in Figures 6 and 7 (Cm
is the velocity component normal to the measuring plane).

In this region, a strong tangential component
Cu appears as may be seen in Figures 6 and 7.

This rotating reverse flow moves up into the suction pipe at a distance L from the impeller inlet depending on the flow rate as shown in Figure 8.

When a guide vane is located in the region where the reverse flow occurs, it takes a mechanical torque Tm equal to a certain fraction of flow momentum. If we consider guiding webs located in the suction pipe, with an internal radius Ri and an outlet radius Ro, the momentum of flow in the corresponding annulus Ro – Ri may be expressed as :

$$M = \int_{Ri}^{Ro} \rho \, 2\pi R^2 \, Cm \, Cu \, dR \qquad (1)$$

ρ = liquid density.

and, when the guiding webs are sufficiently long to cancel the component Cu, we may write :

$$Tm = M \qquad (2)$$

It may then be easily shown that the impeller power input P is increased by a quantity P given by :

$$P = \omega \, Tm \qquad (3)$$

ω = rotational speed of impeller.

The quantities Cm and Cu are functions of the radius R, of the distance L/Do of the considered section and of the relative flow rate $Q* = Q/Q_N$ as shown in Figures 6, 7 and 9.

Figures 9 shows more particularly the variations of Cm and Cu at zero Flow in sections 1 and 3 of Figure 5.

From these Figures, the following explanations concerning the influence of upstream guide vanes can be given :

a/ for a given flow-rate and a given location of guide vanes, the influence of these guide vanes is increasing with the vane width Ro – Ri in accordance with equation (1).

b/ for a given flow-rate and a given width Ro – Ri the influence of the guide vanes is much depending on their relative distance L/Do from the impeller. From Figure 9 it can be seen that, a zero flow, the mean value of Cu near the pipe boundary is decreasing when L/Do increases. In addition the law of variation of Cu with the radium R is also depending on the distance L/Do and reacts on the value of M in equation (1).

5 – PREDICTION METHOD

From the previous results it may be interesting for the designer to predict the approximative shape of performance curves at partial flow for any guiding vane located at any distance L/Do from the impeller. Such a prediction may also be used to choose the right location and dimensions of a guiding web or splitter vane, in order to achieve a guaranteed head Ho or power input Po at zero flow as well as a given slope of the head-flow-rate curve.

5.1 Determination of input power increase at zero flow

It is first necessary to determine the input power increase

ΔPo at zero flow. This quantity may be expressed as :

$$\Delta Po = \rho \, K \, U_1^3 \, R_1^2 \, ((Ro - Ri)/Ro)^a \qquad (4)$$

K is a decreasing function of L/Do as said in the previous chapter.

U_1 : peripheral velocity at impeller inlet

R_1 : impeller inlet radius.

(Ro – Ri)/Ro is the relative width of the guiding vane.

a : is an exponent which is a function of L/Do as may be imagined from results shown in figure 9.

Δ Po has been determined from the values of Po in table 1 by comparing each arrangement with arrangement 1. The corresponding values of L/Do, (Ro – Ri)/Ro, a and K are reported in table 2.

The value a = 1.41 me be easily calculated from the comparison of arrangements 5 and 6 corresponding to L/Do = 2.15. The values of a corresponding to the other values of L/Do have been calculated by assuming the following evolution of a :

$$a = 1.41 + m \, ((L/Do) - 2.15)$$

The value m = – 0.3 has proved to lead to continuously varying values of K as shown in Figure 10 so that relation (4) may be written :

$$\Delta Po = \frac{K(L/Do) \, U_1^3 \, R_1^2}{((Ro-Ri)/Ro)^{1.41-0.3 \, ((L/Do)-2.15)}} \qquad (5)$$

expression in which the values of K(L/Do) are to be red in Figure 10.

5.2 Prediction of the approximate shape of performance curves at partial flow

Basis test results corresponding to a suction duct without any guiding vane are supposed to be available and are represented by curves with a continuous line in figure 11.

The distance L/Do between the impeller and the considered guiding vane leads to a value of Q* drawn form Figure 8. This critical value QK representes the limiting flow-rate under which the performance curves H and Q are influenced by the guiding vanes.

The input power increase ΔPo at zero flow may then be calculated form relation 5 with the help of Figure 10. The corresponding point Po is shown in Figure 11. Then, an approximate input power curve may be drawn between points Po and Pk.

Considering that in all experiments the efficiency curve is nearly unchanged, a new head curve corresponding to the new input power curve Po Pk may be drawn. It is represented by curve Ho Hk in Figure 11.

6 - CONCLUSION

The main scope of this paper is to show that performance curves of a mixed flow pump at partial flow may be adjusted by choosing adequately the dimensions and the location of guiding vanes (webs, splitters or cross). A prediction method based on our experiments is proposed. The field of application of such a method should be enlarged by complementary tests with other specific speed pumps.

REFERENCES :

(1) F. FERRINI - Some aspects of self-induced prerotation in the suction pipe of centrifugal pumps.
WORTHINGTON EUROPEAN TECHNICAL AWARD - PUMPS AND PUMPING SYSTEMS FOR LIQUIDS IN SINGLE OR MULTIPHASE FLOW - Vol. III. HOEPLI-MILANO - 1974.

(2) I. TREBINJAC - Contribution théorique et expérimentale à l'analyse de l'écoulement dans une pompe mixte. Thèse de docteur-ingénieur - INSA - LYON - Nov. 1985.

Table 1

ARRANGEMENT	Ho m	Po kW	$\dfrac{\Delta H}{\Delta Q}$ $\dfrac{m}{m^3/s}$	Qk m^3/s
1	29.0	105	+ 12.2 - 30.8	–
2	30.0	103	+ 8.9 - 32.9	0.040
3	30.6	106	+ 11.25 - 34.7	0.310
4	31.8	108	+ 7.5 - 38.7	0.310
5	31.7	108	+ 6.5 - 32.5	0.200
6	33.6	113	- 21.0	0.200
7	38.0	123.5	- 34.8	0.160
8	40.2	126	- 41.9	0.200
9	42.2	132	- 46.9	0.200
10	42.5	132	- 49.4	0.200

Table 2

ARRANGEMENT	Po (kW)	L/Do	(Ro-Ri)/Ro	a	K x 10³
4	3.0	0.10	0.125	2.03	722.7
5	3.0	2.15	0.129	1.41	191.5
6	8.0	2.15	0.259	1.41	191.5
7	18.5	4.50	1	0.71	65.4
8	21.0	3.09	1	1.13	74.3
9 - 10	27.0	2.60	1	1.28	95.5

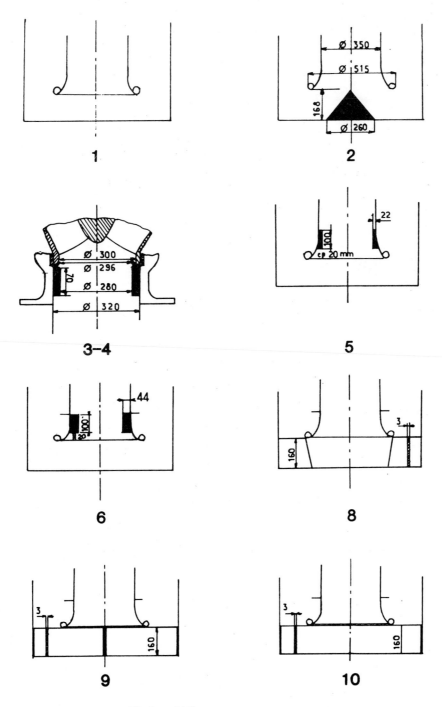

Fig 2   Different test arrangements

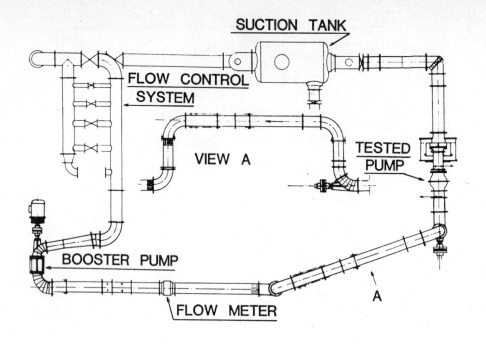

Fig 1   Test loop

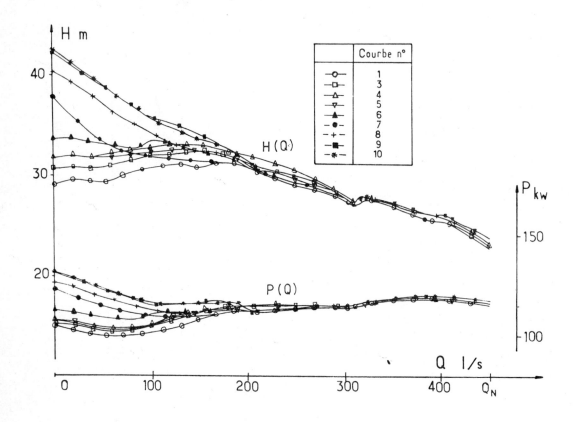

Fig 3   Performance curves

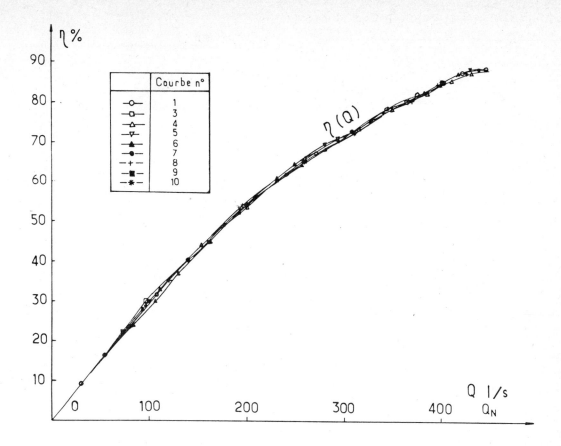

Fig 4   Efficiency curve

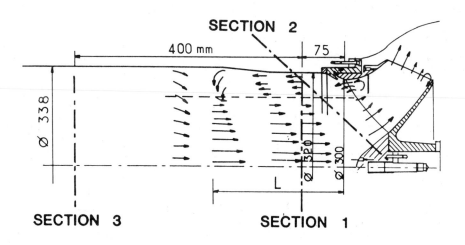

Fig 5   Recirculating flow pattern

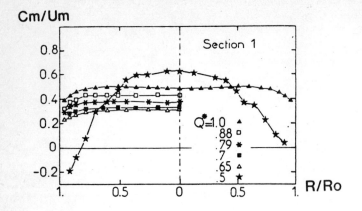

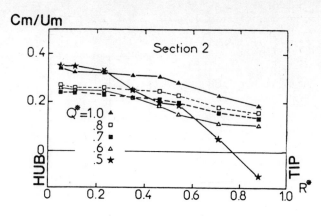

$$R^* = \frac{R - R_{HUB}}{R_{TIP} - R_{HUB}}$$

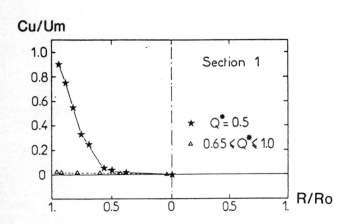

Fig 6   Velocity measurements in section 1

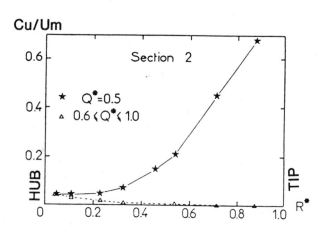

Fig 7   Velocity measurements in section 2

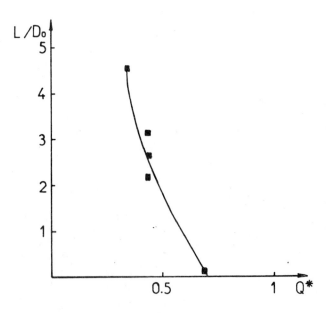

Fig 8   Reverse flow extension

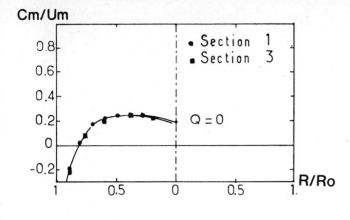

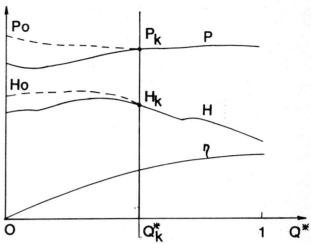

Fig 11   Prediction of performance curves

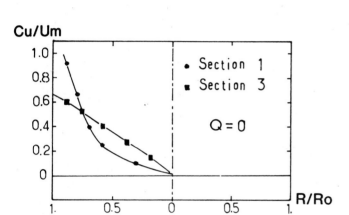

Fig 9   Velocities in section 1 and 3 at zero flow

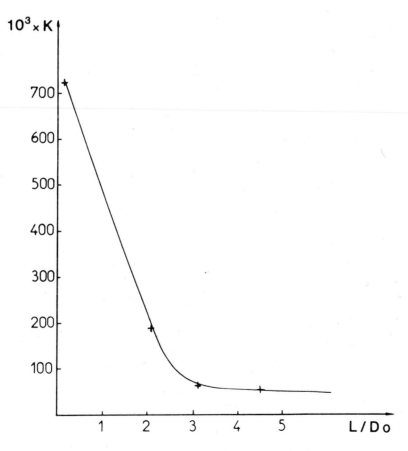

Fig 10   Variation of coefficient K (L/Do)

Partial flow performance comparisons of semi-open and closed centrifugal impellers

A ENGEDA, PhD, MASME and **M RAUTENBERG**, PhD, MASME
Institute of Turbomachinery, University of Hannover, Hannover, West Germany

SYNOPSIS   This paper examines the relative partial flow stage performances of semi-open and closed centrifugal impellers. The impellers were compared on the basis of experimental investigation and reference to literature.

1. NOTATION

b	=	Blade height
C	=	Lift coefficient
d	=	Impeller diameter
H	=	Pump total head
g	=	Gravitational constant
n_s	=	$\omega(Q)^{1/2}/(gH)^{3/4}$, specific speed
P	=	Power input to machine
Q	=	Volume flow rate
U	=	Peripheral velocity
α	=	Incidence angle
δ	=	Tip clearance
λ	=	δ/b, tip clearance ratio
ρ	=	Density
ϕ	=	$Q/2\pi r_2 b_2 U_2$, flow coefficient
ψ	=	$2gH/U_2^2$, head coefficient
ω	=	Angular speed
η	=	$\rho QgH/P$, overall efficiency
μ	=	$2\,P/2\pi b_2 r_2 U_2^3$, Power coefficient

SUBSCRIPTS

Des	=	Design
1	=	At impeller inlet
L	=	Lift
Par	=	Partial
so	=	semi-open
Cl	=	closed

2. INTRODUCTION

Recent trends in pump design are concentrating more and more on the specific questions of: the development of high and variable speed pumps, choice of semi-open or closed impellers, use of inducers, and world wide standardization. A review of available literature shows that the relative performance of semi-open and closed centrifugal pump impellers has up to now received little attention.

Both semi-open and closed impellers are widely used by pump users, but the choice of the semi-open impeller is likely when a pump handling suspended solids or a pump of high specific speed is required. Stepanoff (1) states that the semi-open impeller is more efficient in the medium specific speed range also, this has been confirmed by the present authors (2) in a previous work, which is summarized in figure 1.

The absence of knowledge about the behavior of the semi-open impellers as a function tip clearance and the relative performance of the two type of impellers at partial flow, is the strongest setback for proper choice of impeller type. The occurrence of back flow (inlet and outlet recirculation) and prerotation is accepted to be the determining factor in defining the stability and efficiency of partial flow ranges of pumps. One can easily suspect that because of the differing type of clearance flows and their likely effect on back flow and prerotation, semi-open and closed impellers may have different partial flow performance.

On the basis of experimental investigation of five centrifugal pumps, the partial flow performance of identical semi-open and closed impellers in each pump casing is examined.

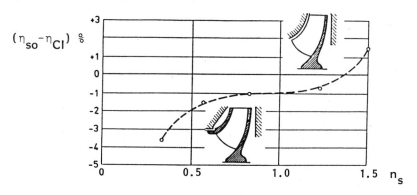

Fig 1   Optimal performance comparison of semi-open and closed impellers as a function of specific speed

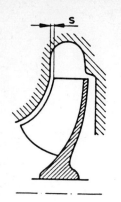 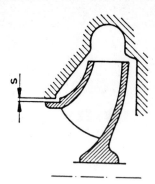

Fig 2   Semi-open and closed impeller

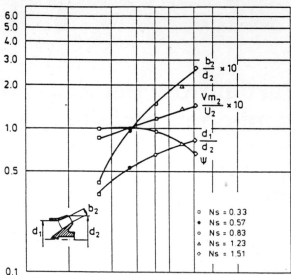

Fig 3   Design characteristics of the tested impellers

3. DESIGN DATA OF TEST IMPELLERS

The five test pumps are typical commercial centrifugal pumps with end suction, volute construction and with closed impellers. The front shroud of the closed impeller was machined off giving a semi-open configuration. The removed front shroud was replaced by an identical shroud which was fastened to the blade tips. For semi-open impeller tests a stationary front shroud with tip clearance variation provisions was mounted. Figure 2 shows the semi-open and closed impeller configuration. The experimental set up consisting of the test pumps, test circuit, measurement technique, and measurement accuracy has been described in detail in (3).

The test pumps were pumps designed according to the same design criteria and they differ only in their specific speeds. Table 1 summarizes the important geometric data of the pumps, and figure 3 shows the design characteristics of the pumps. The pumps were chosen over a wide specific speed range in order to give a good representative picture of the comparison of semi-open and closed impellers.

4. PARTIAL FLOW

Partial flow in centrifugal pumps is characterized by flow incidence variations and the subsequent occurrence of separation, back flow (recirculation), and prerotation, which in turn determines the stability range and efficiency of the machine. In closed impellers there is a leakage through the wearing ring, which adds to the inlet back flow and interacts with the prerotation. In semi-open impellers there is a leakage from the pressure to the suction side of the blade, over the blade tip. This jet leakage has the effect, to unload the blade, to reduce blade effective area,

and to hinder the main flow. Definitely this has, to some degree, an effect on separation and back flow, hence on partial flow performance. However, it can be said that practically, there is no information on this topic both in pump and compressor technology.

Janigro et al. (4) reports from measurements and flow visualization, at the impeller inlet for both semi-open and closed impellers, that prerotation appears to be preceeded by a local separation at the leading vane edge, hinting at the existance of similar type of mechanism.

4.1 Partial Flow Mechanism

The different type of leakage flow in the two type of impellers leads to different behaviors of the impellers with respect to stability and efficiency. Hürlimann's (5) investigation, on a blade cascade into the interaction between flow incidence variation – tip clearance variation – and blade loading showed very interesting results, his conclusions are summarized in figure 4. He was able to show that for a certain flow incidence range, there is little effect on the blade loading at zero clearance (closed impeller case), but as clearance is varied a drastic effect on the blade loading resulted. Thus his results hint at a high dependence of blade performance on the coupled effect of clearance flow and incidence variation. Murakami et al. (6) showed that throttling the inlet of a pump to maintain shockless flow conditions at impeller entrance eliminates back flow, which hinted that incidence has a great effect influence on back flow and prerotation.

n_s	0.33	0.57	0.83	1.25	1.51
D_1	140 mm	145 mm	142 mm	142 mm	180 mm
D_2	404 mm	269 mm	219 mm	185 mm	224 mm
B_2	17 mm	27 mm	32.5 mm	36 mm	59.5 mm
β_1	22°	18°	15°	20°	20°
β_2	26°	24°	26°	28°	25°
Z	6	6	6	6	6

Table 1: Geometrical datas of the impellers

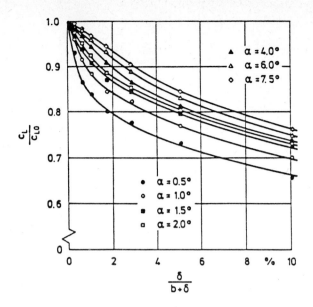

Fig 4   The influence of tip clearance and inlet flow incidence variation
on blade loading, normalized lift coefficient as a function of
clearance ratio and incidence angle [ref (5)]

Harada (7) carried out on the basis of compressor tests, a comparison of the performance of semi-open and closed impellers. He concluded that closed impeller had inferior overall performance to the semi-open impeller at partial flow and that the rotating stall took place at larger flow rates than in the case of the semi-open impeller. His results are summarized in figure 5. Even though his investigation was detailed, he didn't undertake analysis with respect to prerotation and back flow.

An interesting result in that of Kurian et al. (8), who investigated also the relative performance of semi-open and closed impellers and concluded that for the semi-open impeller, the back flow and prerotation began at a lower flow than with the closed impeller. Hinting that the semi-open impeller possesses a better performance at partial flow. Interesting is also the observation of Haupt et al. (9), who on the basis of investigation on a centrifugal compressor, concluded that in a certain range of partial flow, tip clearance flow (in semi-open impeller) is blocked by the back-flow (recirculation) from outlet to inlet along the running edge, again contributing to the improvement of the semi-open impeller's performance at partial flow.

4.2 Present Observation

As discussed above, what little material there is in the literature tends to suggest, that at partial flow the semi-open impeller is likely to posses a better performance characteristic. The present investigation was restricted to stage performance comprisons and with clearance variation in the case of the semi-open impellers. At stage performance level, the relative merits described by Harada (7), Kurian et al. (8) and Haupt et al. (9) for partial flow could not be clearly established.

From the tested five pumps, it was possible to define three categories, in the case of the semi-open impeller: a highly sensitive impeller to clearance loss, a moderate, and an insensitive. Sensitivity is defined on the basis of rate of efficiency change with tip clearance. Figure 6 and 7 show the decrement of efficiency for the most insensitive and sensitive two pumps of specific speed $n_s = 0.57$ and 0.83. For comparison purposes the closed impeller performance, represented by $\lambda = 0$, is also plotted.

When comparing the relative performance of the two types of impellers, the sensitivity criteria was suspected to be an influencing factor of the partial flow performance, but this was found not to be true. The partial flow relative performance of the two type of impellers in fact showed similar behavior in all cases. Figure 8 and 9, as an example, show the general trend of

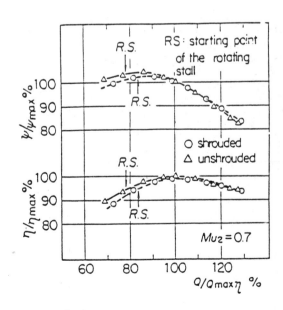

Fig 5   Partial flow performance comparison of a centrifugal compressor, with a semi-open and closed impeller [ref (5)]

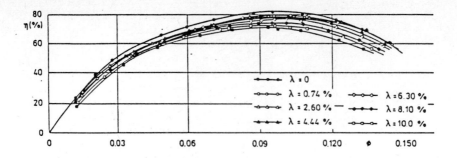

Fig 6   Comparison of semi-open (variable clearance) and closed impeller performance, pump $n_s = 0.57$

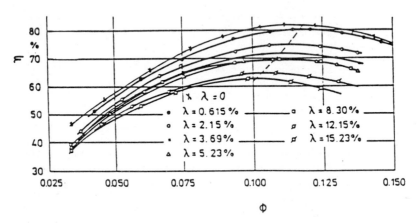

Fig 7   Comparison of semi-open (variable clearance) and closed impeller, pump $n_s = 0.83$

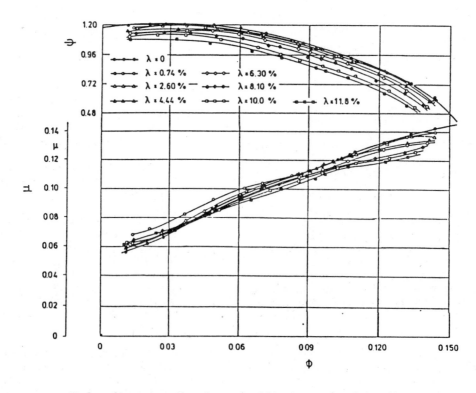

Fig 8   Comparison of semi-open (variable clearance) and closed impeller, pump $n_s = 0.57$

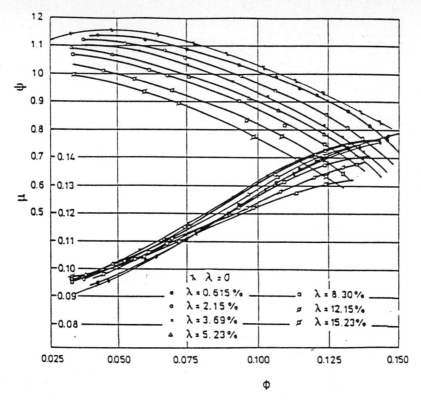

Fig 9   Comparison of semi-open (variable clearance) and closed
impeller performance, pump $n_s = 0.83$

partial flow. In all cases the generated head of the semi-open impellers lies under the closed impeller head, but as flow is reduced to zero, the head generated by the semi-open impellers approached that of the closed, this is probably due to the statement of Haupt et al. (9) that at certain partial flow range back flow blocks clearance flow, saving the kinetic energy loss.

All semi-open impeller showed the tendency of requiring lesser shaft power, as flow is reduced and tip clearance increased, meaning, at partial flow, more clearance flow has the effect of making more blade area ineffective without producing much more loss.

5. CONCLUSIONS

Five pumps of specific speed n_s = 0.33, 0.57, 0.83, 1.25 and 1.51, each with a semi-open and closed impeller were investigated, and the relative stage performance at partial flow was examined. References to literature tend to suggest that the semi-open impeller posses better hydraulic characteristics at partial flow. However, this present work which was restricted to stage performance comparison could not confirm this claim.

It could be concluded that the relative performance at partial flow, judged from stage performances comparisons, is not very much different from the relative performances at design flow.

REFERENCES

(1) Stepanoff, A.J., Centrifugal and Axial Flow Pumps, John Wiley, 1957.

(2) Engeda, A. and Rautenberg, M., The Influence of Specific Speed on the Relative Performance of Half and Fully Shrouded Centrifugal Impellers, 10th Int. Conference of BPMA, Cambridge, 1987.

(3) Engeda, A. and Rautenberg, M., Comparisons of the Relative Effect of Tip Clearance on Centrifugal Impellers, Int. Gas Turbine Conference, 1987, ASME, 87-GT-11.

(4) Janigro, A. and Schiavello, B., Prerotation in Centrifugal Pumps, Von Karman Institute, LS 1978-3.

(5) Hürlimann, R., Untersuchungen über Strömungsvorgänge an Schaufeln in der Nähe von Wänden, Mitteilungen aus dem Institut für Aerodynamic and der ETH in Zürich, Nr. 37, Zürich: Verlag Lehmann, 1963.

(6) Murakami, M. and Heya, N., Swirling Flow in Suction Pipe of Centrifugal Pumps, Bull. ISME, Vol. 9, No. 34, 1966.

(7) Harada, H., Performance Characteristics Shrouded and Unshrouded Impellers of a Centrifugal Compressor, Trl. Engr. Pwr., ASME, Vol. 107, 1985.

(8) Kurian, T. and Rada Krishna, H.C., A Study on the Phenomena of Prerotation at the Suction Side of Centrifugal Pumps, Irrigation and Power, July 1974.

(9) Haupt, U., Chen, Y.N. and Rautenberg, M., On the Nature of Rotating Stall in Centrifugal Compressors with Vaned Diffuser, Part I and II, 1987, Tokyo International Gas Turbine Congress.

C335/88

An idea about the backflow phenomena and their relationship with the internal flow condition of turbomachinery

T TANAKA, MSc, MASME, MJSME, MTSJ, MGTSJ
Department of Mechanical Engineering, Kobe University, Kobe, Japan

SYNOPSIS

It is pointed out in this paper that the traditional expressions about the causes on the occurrence of the backflow phenomena are not sufficient to explain the true backflow phenomena. The reasons behind those erroneous results are discussed. The fundamental causes on the occurrence of the backflow phenomena in a high specific speed turbomachinery may be on its physical structure, that is, on the axis of the rotation of the impeller blades which consists with the flow direction. Because of this, the increasing of the hydraulic resistance on the flow passage of the conduit pipe is recognized as it may act identical to the change of the shape of the flow passages from the open conduit to the closed conduit. Which increases not only the stress which decrease the flow rate, but also the stress which acts similar to the effect of the Centrifugal forces corresponds to the change of the shape of the flow passage.

Both of them may lead at the discharge to make not only the pump head larger, but also the axial velocity and the static pressure head larger at the outer radius and smaller at the inner radius than their averages. They make the flow condition unstable in the radial direction, and may cause the downstream backflow. The inlet flow condition becomes unstable in radial direction because of the pressure gradient which is caused by the deflected flow towards the inner radius and by the deterioration of the blade efficiency after the stall at the periphery. They may act as the major cause on the occurrence of the uptream backflow.

1. INTRODUCTION

In high specific speed turbomachinery, the backflow occurs in the upstream and the downstream of the impeller at off design condition [1,2]. Therefore, the investigation about the backflow phenomena is very significant for the improvement of the reduced efficiency curve at off design condition [3,4]. And it has a very meaningful part, not only in the development of the high specific speed turbomachinery, but also of the low specific speed turbomachinery [5].

In this point of view it seems important to reconsider the traditional hypotheses over again whether they are sufficient to explain the exact reasoning on the occurrence of the backflow phenomena in a turbomachinery, and consider the background of their erroneous results in case of unsufficient. In addition to these, it may be also important to clear the real causes on the occurrence of the backflow phenomena. They are the objectives of this paper and they all are discussed in brief altogether.

2. TRADITIONAL EXPRESSIONS

Stepanoff, A. J. [7] considered that the major cause of the upstream backflow is closely related to the prerotation which is seen in the impeller approach, and he reached to the conclusion that the prerotation is the cause of the upstream backflow and not vice versa. And the downstream backflow is induced by the upstream backflow which was caused previously at a large flow ratio. Professor Murata, S. and Tanaka, S. [2] considered that the upstream backflow is caused by the stalling phenomena of the impeller blades due to the separation of the fluids from the blade surfaces since the angle of attack to the impeller blade becomes large with the decrease in the flow ratio. The upstream backflow region increases with the decrease in the flow ratio and the radial component of the velocity becomes large. Hence the downstream backflow is followingly induced [see Fig.1(a)].

On the other hand, Scheer, W. [3] and Professor Ikui, T. [5] paid their attention to the Centrigugal forces in the flow passage, and considered that as the radial component of the velocity becomes large at the downstream of the impeller at off design condition, the downstream backflow may be caused first at a large flow ratio, and the upstream backflow may be caused later on at a further decreased flow ratio. Professor Toyokura, T. [4] paid his attention also to the Centrifugal force as Scheer and Ikui did. However, his main cause is put on the presence of the casing wall. That is, by the effect of the Centrifugal force the radial component of the velocity becomes large with the decrease in the flow ratio and the fluids start to have a tendency to flow toward the casing wall. This deflected flow may be turned at the casing wall to the axial direction, in most general case to the forward direction. Therefore, the downstream backflow may be caused first at a large flow ratio and the upstream backflow later on at a further decreased flow ratio [see Fig.1(b)].

3. QUESTIONS TO THE TRADITIONAL EXPRESSIONS

Each of those traditional expressions may look very reasonable at a glance since each of their discussions is based upon the practical experimental data. However, there may appear several mental data. However, there may appear several questions when their results are applied to the backflow phenomena which is caused in the opposed order from their own stated order. For example, if the expression by Stepanoff or Murata, which describes the backflow phenomena whose upstream backflow occurs first at a large flow ratio before the occurrence of the downstream backflow, is applied to the backflow phenomena which is seen in the experiments by

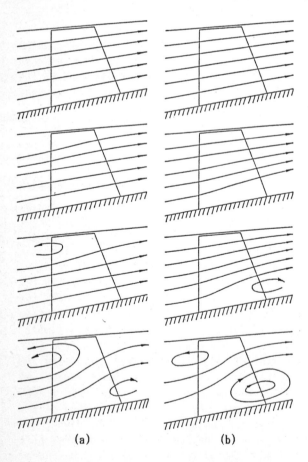

(a)   (b)

Fig 1   Illustrations of the occurrence of the upstream and downstream back-flow phenomena which have been considered

Scheer, Ikui, and Toyokura, it seems not possible to give persuasive reasons on the backflow phenomena. Same problems will be seen when the opposed expression is applied to the backflow phenomena which is caused in the opposed order. In addition to these, when those expressions are applied to the backflow phenomena whose upstream and downstream backflows occur simultaneously at a certain flow ratio, it may be very hard to say that they are extremely reasonable.

4. OBSERVATION OF EXPERIMENTAL TEST RESULTS

The turbomachineries which were used for this investigation are one of the diagonal (semi axial) flow propeller pumps, which has a suction specific speed between those of the mixed flow pumps and the axial flow pumps. Five turbomachineries were examined. Their cross section is shown in Fig. 2. The experimental facilities, equipments, and other detailed performance measurements are reported in reference [6].

The ratios of the flow coefficient at which the upstream backflow was observed to that at the design flow coefficient of DP-0, DP-1, DP-2, DP-3, and DP-4 turbomachineries tested were 0.790, 0.573, 0.603, 0.672, and 0.762, and those at which the downstream backflow was observed were 0.341, 0.653, 0.655, 0.667, and 0.698, respectively.

From the traditional and above results, it could be said that in a practical turbomachinery there are three possibilities in the order on the occurrence of the backflow phenomena. The first case is that the upstream backflow occurs first at a high flow ratio before the occurrence of the downstream backflow. The second case is that the upstream backflow occurs later on at a smaller flow ratio after the downstream backflow occured. And the third case is that both of those upstream and downstream backflows occur simultaneously at some flow ratio.

5. REASONS OF ERRONEOUS RESULTS

If the concluded remarks are kindly allowed to be shown first at the beginning of the discussion, it could be said that there may be four reasons behind the facts that those traditional expressions were not succeeded.

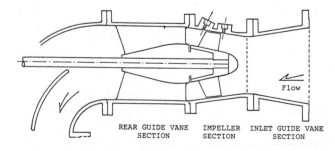

REAR GUIDE VANE   IMPELLER   INLET GUIDE VANE
SECTION   SECTION   SECTION

Fig 2   Cross-section of the turbomachinery tested

The first reason could be found out in the insufficient experimental data which were used for their analysis. Each of them collected the data from the practical experiments. However, it is regretful to find out in their reports that much care was not paied in the order on the occurrence of the backflow phenomena, and that all the orders enclosed are limited to one of those three cases in each investigation. Therefore, the results become insufficient when they meet to the backflow phenomena which are caused in the opposed or in the simultaneous order.

The second reason could be found out in the assumption about the order on the occurrence of the backflow phenomena which is stated individually prior to the analysis. Their order is assumed tacitly in each investigation as the identical to their own experimental order. This would be clear from the discussions in their papers. The cause on the occurrence of the backflow phenomenon which is caused at a large flow ratio is always discussed first and much in detail. But the another backflow which is caused later on at a smaller flow ratio is always discussed afterwards and briefly as if it is caused under some interrelations to the previous backflow. Hence, their results became unsufficient when they meet to the backflow phenomena whose order of the occurrence is in the opposed.

The third reason could be also found out in the discussions. The backflow which is caused at a large flow ratio is always explained as if it induces another backflow later on at a lower flow ratio. This indicates that the flow condition at a small flow ratio is a result affectred by the flow condition at a large flow ratio. However, the continuous change of the flow condition in a flow passage for the decrease in the flow ratio may totally differ from the continuous change of the flow condition in a flow passage at a certain flow ratio whose flow condition between the upstream and the downstream (or the downstream and the upstream) of the blades changes continuously under some dynamic interrelations between them. The former change of the flow condition by the change of the flow ratio may look at a glance in continuous, however they are only the continuous results determined individually and independently to other flow conditions at each flow ratio for the change in the flow ratio following to the character of the shape of the flow passage so that the flow condition might be the best to it at each flow ratio.

Therefore, it is only a continuous view of the changing process of the flow condition at the flow passage for the decrease in the flow ratio. And the flow condition at a certain flow ratio is not the result determined dependently afterwards nor beforehand as a result of some dynamic interrelations with other different flow conditions at a much larger or a much smaller flow ratio. There may exist no common dynamic interrelations between them.

Therefore, it could be said that the flow condition at a certain flow ratio is neither the result which is affected by the other flow condition at a different flow ratio nor the flow condition which affects to another flow condition at some different flow ratio. Hence even though the backflow region at a large flow ratio is large and its development is also large at a certain large flow ratio, it can not become the pure cause of the another backflow which may be caused latter on at a much smaller flow ratio.

The fourth reason could be found out in the understanding about the upstream and the downstream flow conditions for the decrease in the flow ratio. The flow conditions observed in each investigation are the developing processes only either one of the upstream or the downstream backflow. And, they tried to explain the pure causes on the occurrence of the backflow phenomena which has to be explained common to all the backflow phenomena. It was commonly observed in the experiments that the backflow at a large flow ratio increased its flow region for the decrease in the flow ratio. Therefore, the another backflow which is caused later on is explained as if it is induced by the previous backflow as a result of some dynamic interrelations. This expression seems at a glance very reasonable. However, in the DP-O turbomachinery, whilst the flow ratio at which the upstream backflow occured was the largest and the backflow region was the largest and strongest of the five, it was very hard to induce the downstream backflow at a large flow ratio for the decrease in the flow ratio. It was the smallest of the five. There exists, therefore, a question why the flow ratio at which the downstream backflow occured was the smallest. None of the above expressions might be able to give any persuasive reasons about this.

The reason for this may be given as follow. In general, the upstream and the downstream flow conditions at a certain flow ratio are continuous and connected each other with some dynamic interrelations. However, the rate of change for the decrease in the flow ratio are different at the upstream and the downstream even in a turbomachinery. They are of course depend on the shape of the flow passage. Therefore, they may differ if the turbomachineries are different. Therefore, even if the upstream and the downstream flow conditions at a certain large flow ratio are the same (identical) in two of the turbomachineries, there may exist no guarantees that they are still the identical at another different flow ratio. Usually and mostly, they are no longer the identical at other flow ratio. Because even if the upstream and the downstream flow conditions are the identical at a large flow ratio in two turbomachineries, one of them may change the upstream (or the downstream) flow condition so that it growth its flow region at a certain rate for the decrease in the flow ratio, and at the same time it changes the downstream (or the upstream) flow condition at a different rate, and finally at a small flow ratio, for example in the turbomachinery II, it forms a domain of circulation which is totally different from that of the turbomachinery I. Because the rate of change of the flow condition for the decrease in the flow ratio is different at the upstream and the downstream of the impeller when the shape of the flow passage is different. This could be understood from the fact that the back-flow phenomena are caused in the opposed order in the turbomachineries I and II.

The order on the occurrence of the backflow phenomena differs by the character of the shape of the flow passage. Hence, it may not be easy to presume the flow ratio how much has to be decreased after the occurrence of one backflow to the another backflow. In other words, even if the backflow which was caused at a large flow ratio has the most developed and the strongest backflow regions, whenever it may be the up-stream or the downstream backflow, it can not be said that the another backflow would be caused at a relatively small amount of the decreased flow ratio or occurs at a relatively large flow ratio. If circumstances require, lots of the decreased flow ratios may be needed before the occurrence of the another backflow. This would be understood from the experimental facts on the DP-0 and DP-1 turbomachineries.

Now, let us look at the fundamentals on the performance characteristics and the internal flow conditions in a turbomachinery, and discuss the main cause on the occurrence of the upstream and the downstream backflow phenomena.

6. EFFICIENCY CHARACTERISTICS AND BACKFLOW PHENOMENA

The efficiency characteristics of a turbomachinery could be explained in the "efficiency (n) -flow rate (Q)" curve or in the "efficiency (n) - flow coefficient (ϕ)" curve. The former curve expresses the practical efficiency characteristics based upon the revolution number and the latter curve expresses the basic efficiency characteristics regardless to the revolution number. However, these expressions may not be suitable to explain together with all the efficiency characteristics. Because the revolution number, the maximum efficiency, the flow rate (or the flow coefficient) at the maximum efficiency point are different one another among the turbomachineries. The detailed comparison of the efficiency characteristics at off design condition would be reasonable to be done altogether at the identical operating condition, that is at the same flow ratio among the turbomachineries. For this purpose it may be better to rearrange the "efficiency - flow rate" curve

so that the flow condition at the maximum efficiency point is equal to 1.00. That is, the expression "n/n_{max} - Q/Qn_{max}" curve or the "n/n_{max} - $\phi/\phi n_{max}$" curve has to be used.

By the way, the turbomachinery is used to be designed in general so that it demonstrates its maximum efficiency at the design flow rate. And the efficiency deteriorates after passing through the maximum efficiency point with the decrease in the flow ratio. Therefore, it could be said that the "efficiency - flow ratio" curve explains the grade of the deterioration on the efficiency for the decrease in the flow ratio for the whole range of the flow ratios from the maximum efficiency point ($\phi/\phi n_{max} = 1.00$) till the shut-off flow ratio.

If the optimum flow condition at the maximum efficiency point is maintained almost constant for a long period for the decrease in the flow ratio in a turbomachinery, its efficiency might be kept also constant for the long period at a high value, and the resultant shape of the "efficiency - flow ratio" curve may not deteriorate much for the decrease in the flow ratio and it may form a roundish curve with convex. Then, how shall we understand the fluid flow condition at off design condition with such favorable efficiency characteristics in a turbomachinery?

In general speaking, it could be imagined that such turbomachineries may have fully stable flow conditions for the change in the flow ratio. Therefore the change of the velocity distribution may not be performed in extreme, especially at a large flow ratio around the maximum efficiency point. In addition, it could be imagined that the flow condition with the backflow phenomena may totally differ from that without the backflow phenomena which is seen at the high flow ratio around the design flow ratio. Because the former flow condition may be understood rather less stable in a general sence and the fluid flow may involve large hydraulic losses. Therefore, in the turbomachinery which has the favorable efficiency characteristics, the backflow phenomena may not be induced at a large flow ratio, and even if the backflow phenomenon is indused in the flow passage, whether it might be the upstream or the down-stream backflow, the flow ratio at which the backflow occurred may be basically a smaller one. In other words, to perform the "efficiency - flow ratio" curve as a large roundish shape, it could be said that the smaller the flow ratio at which the backflow occurs, whether it might be the upstream or the downstream backflow, it would be desirable. Is this idea really the case? The answer will be shown later on.

6-1. Increasing of Hydraulic Resistance In a Turbomachinery

The flow rate which is produced by turbomachinery may be controlled by the change of the hydraulic resistance μ on the flow passage. If the discharge valve is remained open and the flow rate is controlled by changing the revolution number n, the backflow phenomena may not appear even at a very small flow rate. However, if the flow rate is controlled by keeping the revolution number n constant and the flow rate is decreased by increasing the hydraulic resistance μ on the flow passage of the conduit tube, then the backflow may appear at some off design condition. From these two observations on the experiments it could be said that a change of the flow pattern might not be caused for the former control, but it might be caused for the latter control. From this prospective view at the first stage, it could be understood that the main basic cause on the occurrence of the backflow phenomena seems to be on the increasing of the hydraulic resistance μ in a turbomachinery.

By the way, the increasing of the hydraulic resistance μ on the flow passage is used to be performed by the rise of the hydraulic pressure at the discharge [It may be possible to consider the drop of the hydraulic pressure at the suction instead of the discharge.] or by the closing action of the discharge valve. The former activities do not cause any change in the shape of the flow passage, but the latter activities do. For example, the former activities may be recognized as equivalent to the change of the discharge pipe length (or the suction pipe length) with rough surfaces from zero to infinite. This may be equivalent to the change of the water pressure at the lake where the water is pumping out in the lake (or pumping out from the lake) whose area of the surface is infinite.

The latter activities which operate the valve at the discharge may control the area of the flow passage. Therefore, the shape of the flow passage from the trailing edge of the impelling blades till the discharge valve may vary geometrically with the increasing of the hydraulic resistance μ from the open conduit shape to the closed one, and it forms completely a closed conduit at the shut off flow ratio. That is, the area ratio of the flow passage at the valve to the area of the conduit pipe may vary with the increasing of the hydraulic resistance μ from 1.00, lets say, at the maximum efficiency point till zero at the shut off flow ratio. However, if the discharge valve is assumed to be installed in a location an infinitely long distance from the trailing edge of the impeller, then both of those two activities may become identical regardless whether the discharge valve is established or not.

Therefore, it could be recognized that the increasing of the hydraulic resistance μ on flow passage is identical to the change of the shape of the flow passage from the open conduit to the closed conduit. And this indicates that

the increasing of the hydraulic resistance μ on the flow passage is identical to the increasing, not only the stress which decreases the flow rate, but also the stress which increases a tension similar to the effect of the Centrifugal forces due to the change of the shape of the flow passage. The latter factor is not significant in the flow passage which has no rotating part or which has a rotating part whose axis of rotation is in the vertical to the flow direction, such as in the centrifugal pumps. But it becomes significant when the axis of rotation of the impeller blades is consists to the flow direction, such as in the mixed flow and the axial flow pumps. Because in those flow passages, its effect becomes obvious, especially at off design condition. Therefore, the above recognition may become very significant when we consider the internal flow condition in a high specific speed turbomachinery.

6-2. Decreasing of Flow Ratio

For the fluid flow condition in the range of high flow ratio which is at and near the maximum efficiency point, the ideal flow condition of a perfect turbomachinery, which has the best efficiency at the design condition and the best efficiency curve at off design condition may be assumed. In this case, the mechanical losses, the volumetric losses, and other energy losses could be neglected. Therefore, it would be reasonable to consider that the whole energy E which is supplied to the impeller through the electric motor would be changed whole to the hydraulic energy output by the impelling action. Then the following relation would be obtained.

$$E = f (\gamma Q H) \qquad (1)$$

It could be understood from the general equation (1) that if the impeller is operated at a constant revolution number n for the icreasing of the hydraulic resistance μ, then the flow rate Q would be decreased as $Q_a > Q_b > Q_c > \cdots$ and the head H would be increased as $H_a < H_b < H_c < \cdots$. That is, the following relations may be obtained.

$$\begin{aligned} E &= f (\gamma\, Q_a\, H_a) \\ &= f (\gamma\, Q_b\, H_b) \\ &= f (\gamma\, Q_c\, H_c) \\ &= \cdots \cdots \end{aligned} \qquad (2)$$

Let us consider the change of the flow rate Q and the head H to be produced more in detail. To simplify the discussion, the region of the flow rate Q to be changed is limitted herein as

$$0 \le Q/Q\eta_{max} \le 1.00 \qquad (3)$$

Then the relationship between the energy E supplied and the hydraulic energy output at the maximum efficiency point may be given as follow

$$E_{n\,max} = f(\gamma\; Q_{n\,max}\; H_{n\,max}) \qquad (4)$$

Where the suffix $n\,max$ expresses the physical values at the maximum efficiency point.

At this moment, let us limit the change of the flow condition only in the range of the high flow ratios which is very close to the maximum efficiency point so that the above assumption on the perfect turbomachinery would be practically allowable. Then, the relation of equation (4) may be applicable to the flow condition which is very close to the maximum efficiency point.

Remaining the energy supply $E_{n\,max}$ at a constant, if the flow rate Q_o is reduced from the flow rate $Q_{n\,max}$ at the maximum efficiency point as much as the very small amount $\triangle Q_{o1}$, then the flow rate Q_{o1} becomes as follow

$$Q_{o1} = Q_{n\,max} - \triangle Q_{o1} \qquad (5)$$

In the perfect turbomachinery the energy which is equivalent to the reduced flow rate $\triangle Q_{o1}$ may be used whole to the production of the head $\triangle H_{o1}$. Therefore, the head H_{o1} at the flow rate Q_{o1} may be obtained as follow.

$$H_{o1} = H_{n\,max} + \triangle H_{o1} \qquad (6)$$

Then, the following relations would be obtained for various flow rates at and near the maximum efficiency point.

$$
\begin{aligned}
E_{n\,max} &= f(\gamma\; Q_{n\,max}\; H_{n\,max}) \\
&= f[\gamma\;(Q_{n\,max}-\triangle Q_{o1})(H_{n\,max}+\triangle H_{o1})] \\
&= f[\gamma\;(Q_{n\,max}-\triangle Q_{o2})(H_{n\,max}+\triangle H_{o2})] \\
&= f[\gamma\;(Q_{n\,max}-\triangle Q_{o3})(H_{n\,max}+\triangle H_{o3})] \\
&= \cdots\cdots \qquad (7)
\end{aligned}
$$

Therefore, it could be understood that the head H_o produced always becomes much higher than that at the design flow rate. That is,

$$H_{n\,max} < H_o \qquad (8)$$

And its difference becomes large with the decrease in the flow rate. That is, as $Q_{n\,max} > \triangle Q_{o1} > \triangle Q_{o2} > \triangle Q_{o3} > \cdots$,

$$H_{n\,max} < \triangle H_{o1} < \triangle H_{o2} < \triangle H_{o3} < \cdots \quad (9)$$

Therefore, the shape of the "head - flow rate" curve of all the turbomachineries has a fundamental gradient which is fallen down to the right for the change in the flow ratio. In other words, if the change of the flow rate Q_o at off design condition is limitted in the range very close to the design flow rate, [note: Above relation may be satisfied even at the range larger than that at the design flow rate. However, the discussion about it is neglected here to simplify the discussion.] then the axial pressure gradient from the impeller inlet to the discharge becomes larger than that at the design flow rate and it becomes much larger with the decrease in the flow rate.

On the other hand, if the hydraulic resistance μ on the flow passage is increased in a turbomachinery, then the flow rate decreases. And the optimum flow condition for the flow passage, that is the optimum flow pattern may be produced at each flow rate. The character of the flow pattern at each flow ratio may not be changed by the revolution number in a turbomachinery because the law of similarity is satisfied if the flow ratio is remained at a constant value. However, if the turbomachineries are different, even if the flow ratio is the same, the character of the flow pattern may differ one another among the turbomachineries since the shape of the flow passage differs among them.

The amount of the energy which is supplied to the impelling blades may not be constant in a practical turbomachinery even though the operating revolution number is remained constant. As it is seen in the "power - flow ratio" curve, it increases with the decrease in the flow ratio in a practical high specific speed turbomachinery, see Fig. 2. Its grade and the character may differ if the shape of the flow passage differs. [note: This fact is well known among the turbomachinery engineers. However, it seems that this fact is not concerned much in the current theoretical investigations.] The "power - flow ratio" curve at off design condition on the practical turbomachinery could be explained as that it indicates the amount of the energy consumption at each flow ratio to obtain the same revolution number as it was at the maximum efficiency point. That is, the amount of the energy consumption per unit revolution number at each flow ratio may be explained in the "power - flow ratio" curve for the whole range of the flow ratios. As the efficiency characteristics at off design condition of a turbomachinery deteriorates gradually with the decrease in the flow ratio from the maximum efficiency point to the shut-off flow rate, and also the gradient of the "power - flow ratio" curve is fallen down to the right, the change of the flow ratio indicates that the flow pattern changes from the flow pattern whose energy consumption per unit revolution number is economical to the less economical one without any change in the shape of the flow passage.

The supplemental energy $\triangle E_{a1}$ at the flow rate Q_{o1} which is supplied additionally to obtain the same revolution number as it was at the design flow ratio may be also displaced whole to the hydraulic head energy $\triangle H_{a1}$ by the impelling action in the perfect turbomachinery. And its amount would be added to the head H_{o1} which is obtained at the flow rate Q_{o1} by the energy displacement of the fundamental energy $E_{n\,max}$

which is supplied at the design flow ratio. That is, the energy E_1 which is practically supplied at the flow rate Q_{o1} may be given by

$$E_1 = E\eta_{max} + \triangle E_{a1}, \qquad (10)$$

and hence it may be explained in the form of a general equation, as

$$E = E\eta_{max} + \triangle E_a \qquad (11)$$

And the practical head H_1 which is produced at the flow rate Q_{o1} may be given as follow.

$$\begin{aligned}H_1 &= H_{o1} + \triangle H_{a1} \\ &= H\eta_{max} + \triangle H_{o1} + \triangle H_{a1}\end{aligned} \qquad (12)$$

and again it may be given in the form of a general equation as follow.

$$H = H\eta_{max} + \triangle H_o + \triangle H_a \qquad (13)$$

As described above, the amount of supplemental energy $\triangle E_a$ may increase with the decrease in the flow ratio in the practical turbomachinery. That is, as $Q\eta_{max} > \triangle Q_{o1} > \triangle Q_{a2} > \triangle Q_{o3} > \cdots$,

$$\triangle E_{a1} < \triangle E_{a2} < \triangle E_{a3} < \cdots \qquad (14)$$

Therefore, the amount of the supplemental head $\triangle H_a$ which is obtained by the energy displacement of the supplemental energy $\triangle E_a$ increases as

$$\triangle H_{a1} < \triangle H_{a2} < \triangle H_{a3} < \cdots \qquad (15)$$

Consequently, as the head $\triangle H_o$ by the fundamental energy displacement and the head $\triangle H_a$ by the supplemental energy displacement in equation (13) are always positive at the off design condition, the head H which is practically produced at the off design condition may be larger than that at the maximum efficiency point. That is,

$$H\eta_{max} < H. \qquad (16)$$

And, as the magnitudes of $\triangle H_o$ and $\triangle H_a$ in equation (13) increase with the decrease in the flow ratio, the equation (16) may be explained as follow.

$$H\eta_{max} < H_1 < H_2 < H_3 < \cdots \qquad (17)$$

Therefore, in the practical turbomachinery the axial pressure gradient from the entrance to the discharge of the impeller may become much larger at off design condition than that at the design flow rate and its gradient becomes much larger with the decrease in the flow ratio than that which is obtained from the assumption that the amount of the energy supply is equal to that at the maximum efficiency point.

6-3. Occurrence of Downstream Backflow

From the above discussions it would be clear that the increasing of the hydraulic resistance μ on the flow passage is identical to the increasing, not only the stress which decreases the flow rate, but also the stress which increases a tension similar to the effect of the Centrifugal forces due to the change of the shape of the flow passage although the shape of the flow passage is not necessary to be changed by the increasing of the hydraulic resistance μ on the flow passage in a practice. The former stress may act to increase not only the attack angle, but also the axial pressure gradient between the impeller inlet and the outlet sections (pump head) large, that is the pressure head large at the discharge at off design condition. And the second stress may act to cause the axial velocity and the pressure head higher at the outer radius and lower at the inner radius at off design condition.

In other words, in a practical turbomachinery they may act at the discharge to make not only the head, but also the axial velocity larger at the outer radius and smaller at the inner radius than their averadges. That is, the fluid flow at the impeller discharge becomes less stable in the radial direction. And their unstable grade becomes much larger with the decrease in the flow ratio. Therefore, the head and the axial velocity distributions on the radius at off design condition for the decrease in the flow ratio have some meaningful significance on the occurrence of the downstream backflow phenomenon in a turbomachinery.

From the above observations, it could be concluded that the fundamental cause on the occurrence of the downstream backflow may be on its physical structure of the high specific speed turbomachinery, that is, on the axis of rotation of the impeller blades which consists with the flow direction. Because of this, the increasing of the hydraulic resistance μ on the flow passage acts to increases not only the stress which decrease the flow rate, but also the stress which increase a tension similar to the effect of the Centrifugal forces which corresponds to the change of the shape of the flow passage. And they leads at the discharge to make not only the head, but also the axial velocity larger at the outer radius and smaller at the inner radius than their averadges. They make the flow condition unstable in the the radial directions, and cause the downstream back flow in the outlet of the impeller.

6-4. Occurrence Of Upstream Backflow

In the practical turbomachinery which is operated at a constant revolution number, if the hydraulic resistance μ on the flow passage is increased, it may act not only to make the flow ratio decrease, but also the head and the axial velocity larger at the outer radius and smaller at the inner radius than their averadges at the discharge. The former acts to make the attack angle large. And the latter acts to make the pressure and the axial velocity gradients between the impeller inlet and the outlet larger at the outer radius and smaller at the inner radius than that which is obtained by the equations (16) and (17). The discharge head, therefore, becomes much larger than the pumping ability of the blade elements at the outer radius, but smaller at the inner radius. This may act to decrease the inlet axial velocity at the outer radius and increase at the inner radius. These may lead to make the attack angle larger at the outer radius and smaller at the inner radius. Therefore, it could be said that both of them may act to make the attack angle larger at the outer radius and smaller at the inner radius for the approaching inlet flow.

In addition to this, the blade angle is the smallest at the periphery of the leading edge. As it is seen in the experimental test results by Professor Ikui, the boundary layer is thicker at the periphery because of the effects of the Centrifugal force and the secondary flows along the blade surfaces. And the inlet flow condition is less stable there. Therefore, the fluid flow at the periphery of the leading edge is under the flow condition very easy to induce the separation of the fluids from the blade surfaces and to make the stall of the blades for the increased attack angle. These consists with the results pointed out by Scheer and Murata [2,3].

By the way if we look at the experimental test results on the practical turbomachineries such as of the DP-0 turbomachinery which induced the upstream backflow first at a large flow ratio and of the DP-1 turbomachinery which induced the downstream backflow first, although their orders on the occurrence of the upstream and the downstream backflows are in the opposed, it would be observed common to both turbomachineries that the blade efficiency is dropped rapidly at a large flow ratio before the occurrence of the upstream backflow and that the flow ratio is fairly larger than that at which the upstream backflow occured. The grade of the efficiency drop is very large and hence very clear to realize in the "efficiency - flow ratio" curve, as it is seen in Fig. 3. This indicates that as the attack angle increases with the increase in the hydraulic resistance μ on the flow passage, the tension which maintained its flow condition optimum for the decrease in the flow ratio had reached to its limit at the flow ratio, and caused the separation of the fluids and the stall of the blades. And hence the blade efficiency dropped rapidly.

In other words, it could be said that the upstream backflow occured at a fairly decreased flow ratio than the stalling point at which the blade efficiency dropped rapidly in those turbomachineries. This indicates that the upstream backflow may not be induced directly or immediately after the efficiency drop was caused by the increasing of the attack angle, that is by the separation of the fluids from the blade surface and the stall of the blades. This also indicates that it was very necessary for the interanl flow condition to decrease the flow ratio to construct a deteriorate unstable flow condition which is sufficient and good enough to produce the upstream backflow phenomenon in the flow passage. This understanding on the experimental facts would be very significant to understnad the causes on the occurrence of the upstream backflow. This also indicates that the qualification on the occurrence of the upstream backflow by Professor Murata and Tanaka is still not sufficient.

In the meantime, it could be understood that in a turbomachinery the efficiency drop may be caused by the unstable flow condition on the blade surfaces. The unstable flow condition might be caused by the separation of the fluids from the blade surfaces and by the stall of the blades in the field of two dimensional cylindri-

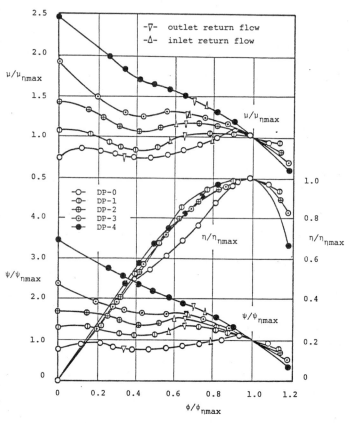

Fig 3   General performance characteristics of turbo-machineries tested and the flow ratios at which the upstream and downstream back-flows started (symbols ϕ, ψ, μ designate measured flow, head, power coefficients, η overall efficiency in percentage terms, and $\phi_{\eta}\text{max}$, $\psi_{\eta}\text{max}$, $\phi_{\eta}\text{max}$, $\psi_{\eta}\text{max}$ those coefficients at the maximum efficiency point)

© IMechE 1988 C335/88

cal surface and its effect may appear in the "efficiency - flow ratio" curve as the drop of the efficiency at a fairly large flow ratio before the occurrence of the upstream backflow. It seems, therefore, reasonable to consider that the two dimensional effect is deeply related to the drop of the overall efficiency in the turbomachinery. It seems, however, that it may rather acts as an indirect part on the occurrence of the upstream backflow phenomenon since the upstream backflow is induced at a fairly small flow ratio than that at which the efficiency drop was caused rapidly. In addition, the upstream backflow involves the radial velocity component. It seems, therefore, not easy to explain the pure cause on the occurrence of the upstream backflow phenomenon only by discussing the unstable factors which appear in the two dimensional cylindrical flow field since it does not contain the radial component.

As the cause on the occurrence of the upstream backflow, the authour would like to point out the discharge head which becomes larger than the pumping ability of the blade elements at the outer radius. This may act to make the inlet velocity smaller at the outer radius and larger at the inner radius. The discharge head becomes especially larger (the largest) at the periphery which, therefore, makes not only the attack angle rapidly large, but also the pressure difference between the impeller inlet and the outlet large rapidly at the periphery. For the less favoravle turbomachinery this may act to cause the backflow phenomenon towards the upstream of the impeller. The other cause could be found out on the unstable flow condition in radial direction due to the pressure gradient which is caused by the deflected flow torwards the inner radius at the impeller inlet and the deterioration of the blade efficiency on the energy displacement after the stall at the periphery. And the upstream backflow, therefore, may be caused after those unstable flow conditions are satisfied in the axial and the radial directions.

Furthermore, the order on the occurrence of the backflow phenomena could be defined due to the relative grade of the upstream and the downstream flow conditions in a turbomachinery.

7. RESULTS

It was pointed out in this paper that the traditional expressions about the causes on the occurrence of the backflow phenomena are not sufficient to explain the true backflow phenomena. The reasons behind those erroneous results were discussed. The fundamental causes on the occurrence of the backflow phenomena in a high specific speed turbomachinery may be on its physical structure, that is, on the axis of the rotation of the impeller blades which consists

with the flow direction. Because of this, the increasing of the hydraulic resistance on the flow passage of the conduit pipe is recognized as it may act identical to the change of the shape of the flow passages from the open conduit to the closed conduit. Which increases not only the stress which decrease the flow rate, but also the stress which acts similar to the effect of the Centrifugal forces corresponds to the change of the shape of the flow passage.

Both of them may lead at the discharge to make not only the pump head larger, but also the axial velocity and the static pressure head larger at the outer radius and smaller at the inner radius than their averages. They make the flow condition unstable in radial direction, and may cause the downstream backflow. On the other hand, such downstream flow condition may affect to the upstream condition. At the periphery the discharge head becomes especially larger than the pumping ability of the blade elements, which makes the inlet velocity smaller at the outer radius and larger at the inner radius. The pressure gradient due to the deterioration of the blade efficiencies after the stall may be also significant. They may act as the major cause on the occurrence of the uptream backflow.

REFERENCES

1. Tanaka, T.,"An Evaluation of Efficiency Characteristics based on Internal Flow Condition of Pumps," Small Hydro Power Fluid Machinery-1982, ASME, November 1982, pp.67-71.
2. Tanaka, S. and Murata, S.," On the Partial Flow Rate Characteristics of Axial Flow Compressor and Rotating Stall, 1st and 2nd Reports", Trans. JSME, Part 2, Vol.40, No. 335, July 1974, pp.1938-1947 and 1948-1957.
3. Scheer, W., "Untersuchungen und Beobachtungen uber die Arbeitsweise von Axialpumpen unter besonderer Berucksichtigung des Teillastbereiches", BWK, Kraft Bd.11, Nr.11, November 1959, pp.503-511.
4. Toyokura, T., Kitamura, N., and Kida, K., "Studies on the Improvement of High-Specific-Speed Pump Performance in the Domain of Low Flow Rates," Journal of JSME, Vol.67, No.544, May 1964, pp.682-691.
5. Ikui, T., "Flow Conditions in the Axial Flow Blades", Journal of JSME, Vol.62, No.485, June 1959, pp.909-918.
6. Tanaka, T., " An Experimental Study of Backflow Phenomena in a High Specific Speed Turbomachinery", Proceedings of the 10th BPMA International Pump Technical Conference "The Pressure to Change", Churchill Colledge, Cambridge, England, March 24-26, 1987, BHRA Fluids Engineering, pp.41-60.
7. Stepanoff, A. J., Pumps and Blowers: Two Phase Flow, John Wiley and Sons, Inc., New York, N. Y., 1965, pp.83-95.

C336/88

The flow field at the tip of an impeller at off-design conditions

A GOULAS, Dipl-Ing, MSc, DIC, PhD
Laboratory of Fluid Mechanics, University of Thessaloniki, Thessaloniki, Greece
G TRUSCOTT
Formerly BHRA, Cranfield, Bedfordshire

ABSTRACT

The flow at the exit of the impeller of a centrifugal pump was measured at various positions relative to the cut-water using a Laser Doppler velocimeter.

The flow at the exit is very distorted. The degree of distortion is a function of flow-rate and cut-water design.

Away from cut-water a reverse flow region was observed near the back shroud. This diminishes and finally disappears when the flow rate increases.

Close to cut-water at low flow-rates there is a radial inward flow. The area of such flow again is reduced increasing the flow rate.

Past the cut-water and for low flow-rates a reverse flow region was observed.

Finally in the same region at low flow-rates there is a double vortex structure, secondary flow, while at large flow rates.

1. INTRODUCTION

The flow field at the exit of a centrifugal pump impeller is a good indication of the flow mechanisms within the impeller on one hand, and of the interaction between volute flow and impeller on the other.

This is the reason for the extensive literature on the velocity profiles at the exit of impellers. Because of the highly distorted flow at exit, rotating blade wakes etc, there has been a need to measure those velocity profiles both non-invasivally and with an instrument of high frequency response.

For pumps, hot-film and laser doppler velocimeters (LDV) are the two most promising techniques.

In the case of Laser doppler velocimetry there are several publications reporting measurements in or around pump impellers. Howard et al (1) measured the flow within and at the exit of a centrifugal pump impeller with helical blades and for three flow conditions. Later Hamkins (2) reported measurements on a fully transparent radial pump impeller. Peacock et al (3) reported measurements at the inlet of a centrifugal pump while Fraser et al (4) and Radke et al (5) reported detailed velocity measurements in a mixed flow pump. Laser doppler velocimeter measurements inside a rotating axial flow pump were reported by Resnick et al (6).

All the above measurements high-lighted various aspects of the flow around the impeller, but there was no indication about the interaction effects between volute and impeller outlet flow.?
In this work, detailed measurements of the flow at various flow conditions close to and away from the cut-water will be reported.

2. EXPERIMENTAL APPARATUS

The experimental apparatus consisted of an end-suction volute pump (150/150 mm), specific speed (SI units) 0,685 and the appropriate flow circuit. On the front shroud of the pump, perspex windows were fixed in order to facilitate optical access for the laser beams. Details of the pump are given in table 1.

Fig. 1 shows the meridional section of the impeller.

The LDV consisted of a single component back scatter system. The processing system was based on an externally driven spectrum analyser. The frequency/velocity information together with a signal related to the position of a blade at the time of the velocity measurement was stored into a computer for further processing. By turning the system through 90° two components of velocity could be measured, radial and circumferential, together with the corresponding RMS values.

The instantaneous values of velocity were sorted out according to their circumferential position. For each position then the mean value:

$$\bar{V}_x = \Sigma \, V_i / N$$

and the RMS:

$$\bar{V}_x^{\,2} = \frac{1}{N-1} \Sigma \, (V_i - \bar{V})^2$$

where x stands for either r (radial) or u (circumferential), was calculated.

For each position the size of sample N was between 200 and 400. The statistical error for the turbulence intensities encountered was estimated between 5% and 3% respectively.

The majority of measurements were taken at a radial position of $D_2/2+5$ mm and at five axial positions.

In order to study the effect of cut water on the flow coming out of the impeller measurements were taken at eight circumferential positions seven close to cut-water, between -16° to 52° from the radial plane through the cut water, and one away from it at 260°.

Five flow-rates were studied. The exact values are given in Table 2.

In the next section the results will be presented and discussed.

3.   **RESULTS AND DISCUSSION**

Fig. 2 shows the H-Q characteristic of the pump and the efficiency - flow-rate curve. Although the nominal design point was at $188m^3$/h the measurements indicated that the BEP was at $248m^3$/h. The percentages of flow-rate relative to the last value are given in brackets in Table 2 and are going to be used in this report. The pump is an old design and has relatively low efficiency. This may be related to the distorted profiles observed at the exit of the impeller.

Figs. 3a to 3c show the radial velocity profiles, and 4a to 4c the circumferential absolute velocity profiles at the exit of the impeller away from the cut water for the following flow-rates 55,6%, 103% and 123%.

The results are plotted in the form of iso-velocity contours. In the case of

the radial velocity plots the data were non-dimensionalised with the average axial velocity at the inlet of the pump.

For the circumferential velocities the results were non-dimensionalised using the tip velocity $u_2 = \omega \, r_2$.

At the 55,6% flow-rate there is a clear jet-wake structure in the radial velocity profile with high velocity near the pressure side and a wake in the corner of suction side/front shroud.

Increasing the flow-rate the structure of the flow at the exit changes Fig. 3b.

A reverse flow region develops at the back shroud while close to the front shroud the velocity is more uniform.

The reverse flow region covers the whole circumference, and is similar to that observed by Mizuki et al (7) in compressors. Increasing the flow-rate the reverse flow region disappears firstly near the pressure side, and finally disappears completely at a flow-rate of 123% BEP. Nevertheless the region close to the back shroud is still a low flow, wake, region.

The explanation of why there is such a change in velocity profiles at low flow-rates may lie in the inlet velocity profiles of the same pump as measured by Peacock et al (3).

At the 50% BEP it was found that a reverse flow region was developed at inlet and it was coupled with strong swirl. The high velocities observed near the back shroud together with the swirl may be responsible for the destruction of the low flow region at the back of the impeller and the formation of wake at the suction side/front shroud corner.

The circumferential components of the absolute velocity show a minimum at the region of high radial velocity, jet side, and a maximum at the low radial velocity region. Furthermore the reverse flow core observed at the back shroud region has a constant Vu equal to 0,4 of the rotational speed at the tip.

The above measurements showed a high degree of flow distortion near the exit of the impeller. The measurements close to the cut-water, however, showed a very different picture. The blade-to-blade variation of the flow was minimal compared to that away from the cut-water; and reinforced the opinion that the cut-water imposed its own velocity distribution on the impeller in this region. Measurements of the pressure distribution around the volute reported by Goss (8), showed a very strong circumferential pressure gradient near the cut-water which obviously affected the flow coming out of the impeller.

Fig. 5a shows the velocity vectors near the cut-water for the low flow-rate (27% of BEP). The flow is towards the cut-water clearance indicating a low static pressure region. This finding was supported by visual observation of cavitation in the region of cut-water clearance. The region of radial inward flow (negative V_r) extends upstream of the cut-water.

Fig. 5b, shows the velocity vectors near the cut-water for the high flow-rate 123% of BEP. The flow is pointing away from the cut-water indicating a possible separation at the outside of the cut-water.

Figs. 6a to 6c show the variation of radial velocity at a fixed impeller position (middle passage close to front shroud) as a function of the angular position from the cut-water. At low flow-rates the region of inward flow extends from −15° to 10°. Increasing the flow-rate (Fig. 6b) the inward flow region is reduced to areas upstream of the cut-water. At BEP and flow rates above it (Fig. 6c) the radial velocity is almost constant.

Fig. 7 shows the velocity vectors downstream of the cut-water and for a flow rate 28% of BEP. A reverse flow region is indicated from the measurements.

Finally in Fig. 8, the circumferential components of the absolute velocity are shown as a function of the distance between front (F.S.) and back (B.S.) shroud and for all five flow-rates. The measuring position is 19° away from the cut-water plan.

For flow-rates below BEP, case 1 to 3, the V_u has a maximum within the passage indicating a double vortex structure, secondary flow, at the exit of the impeller. At the low flow-rate the two vortices appear to be of the same size, symmetrical. Increasing the flow-rate the vortex close to the back shroud increases in strength, steeper velocity gradient in the axial direction, and covers a larger area. At the BEP and for flow-rates above it, cases 4 and 5, the back vortex dominates the flow-field while its strength increases.

4.   CONCLUSIONS

The flow at the tip of the impeller of a centrigugal pump is very distorted. The degree of distortion depends mainly on the flow-rate and the position relative to the cut-water.

Away from the cut-water at low flow-rates there is a jet-wake structure of the flow. Around BEP there is a reverse flow region at the back shroud. The extent of this region is a function of the flow-rate.

Close to the cut-water the flow is dominated by the pressure field imposed by the cut-water, there is a region of inward flow at low flowrates. Increase in flow reduced the area of inward flow while close to BEP the radial velocity is positive and constand around the cut-water.

Past the cut-water there is a reverse flow region at low flow-rates.

Finally, in the same region for low flow-rates there is indication of a double vortex structure coming out of the impeller.

At flow-rates close to, and above BEP, there is only a single vortex moving in the same manner as the back shroud vortex observed at low flow-rates.

5.   ACKNOWLEDGEMENTS

This work was supported by five pump manufacturers, Byron Jackson Holland, Hayward Tyler Ltd., Sultzer Bros. (U.K.) Ltd., Weir Pumps Ltd., Worthington-Simpson Ltd. and the DTI.

The authors would like to thank BHRA, The Fluid Engineering Centre and Mr. M. Cornell for their help and advice during the running of the project.

REFERENCES

1.   HOWARD J.H.G. et al
"Laser Doppler measurements in a radial pump impeller".
ASME Gas Turbine Conference,
New Orleans 1980.

2.   HAMKINS C.P.
"Laser velocimeter measurements in a radial pump impeller"
M.A. Thesis Univ. of Virginia, 1985

3.   PEACOCK G., GOULAS A.,
"LDA measurements of the flow at the inlet of a centrifugal pump"
BPMA 9th Tech Conference Paper 7, 1985

4.   FRASER S.M., CAREY C.,
"Laser Doppler measurements in the region between the rotor and the stator of a mixed-flow pump"
BPMA 8th Technical Conference, 1983

5.   RADKE M., SCHRODER R., SIEKMANN H.E.,
"Application of Laser Doppler velocimetry to a fluid mechanics investigation of a mixed flow tubular casing pump"
KSB Technische Berichte No 20e, 1986

6.   RESNICK A., GOULAS A.,
"Laser Doppler measurements inside an axial pump impeller at off-design conditions"
IAHR 11th symposium on hydraulic machinery Amsterdam 1982

7. MIZUKI S., et al,
   "Reversed flow phenomena within centrifugal compressor channels at low flow-rates"
   ASME 76-GT-86, 1986

8. GOSS M.
   "Fluctuating thrusts in centrifugal pumps"
   BHRA CR1606, 1979

AP/aca/AP-REPORT/BGC/27.04.88

Table 1 — Pump details

Details of Test Pump (nominal dimensions)	
Nominal Duty	: 191m^3/h (700 GPM) and 19,2m (63 ft) head at 1450 rev/n
Branch Sizes	: 150 mm (6´´)
Impeller OD	: 254 mm (10´´) maximum
Impeller ID	: 133 mm (5,25´´)
Impeller outlet width	: 28,5mm(1,125´´)(+12,5mm(0,5´´) for two shrouds)
Number of blades	: 6
Nominal outlet angle	: 35°
Nominal inlet angle	: 30°
Volute form	: rectangular, with fillets
Volute throat area	: 9290 mm^2 (14,4 ins^2)
Volute width	: 82,5 mm (3,25´´)
Cut-water clearance	: 4mm (0,16´´)nominal, 3mm (0,125´´) actual
Wear ring diameter	: 149,0mm (5,868´´)front, 149,7mm (5,893´´)rear.

Table 2

Code No	FLOWRATE m^3/h	PERCENTAGE OF BEP
1	70,6	37% (29,4)
2	134	70% (55,6)
3	198	105% (82,5)
4	248	129,6% (103)
5	296	155% (123)

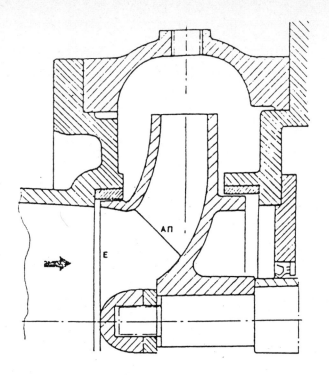

Fig 1   Cross-section of the impeller

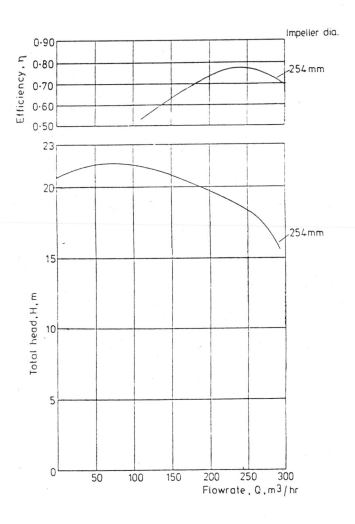

Fig 2   H—Q characteristics

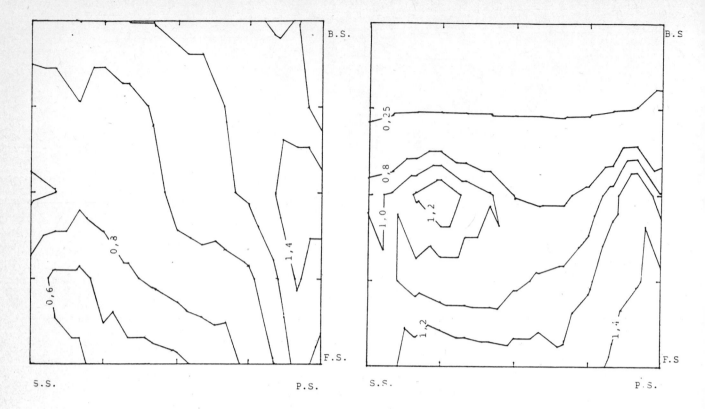

Fig 3a   Iso-velocity contours of the radial velocity at exit
         for 55.3 per cent BEP

Fig 3b   Iso-velocity contours of the radial velocity at exit
         for 103 per cent BEP

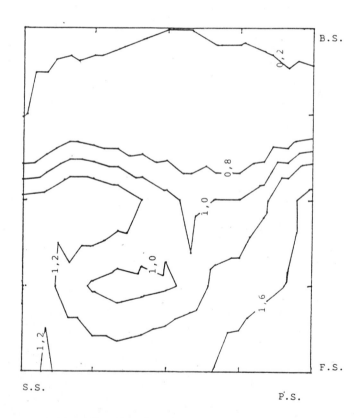

Fig 3c   Iso-velocity contours of the radial velocity at exit
         for 123 per cent BEP

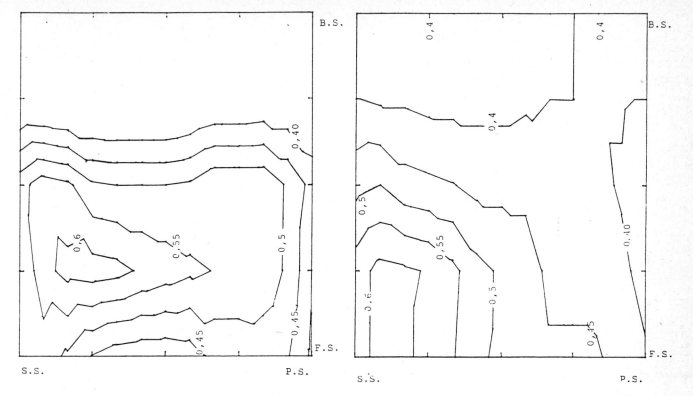

Fig 4a Iso-velocity contours of the circumferential velocity at the exit for 55.3 per cent BEP

Fig 4b Iso-velocity contours of the circumferential velocity at the exit for 103 per cent BEP

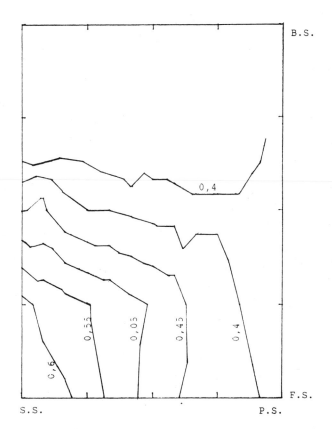

Fig 4c Iso-velocity contours of the circumferential velocity at the exit for 123 per cent BEP

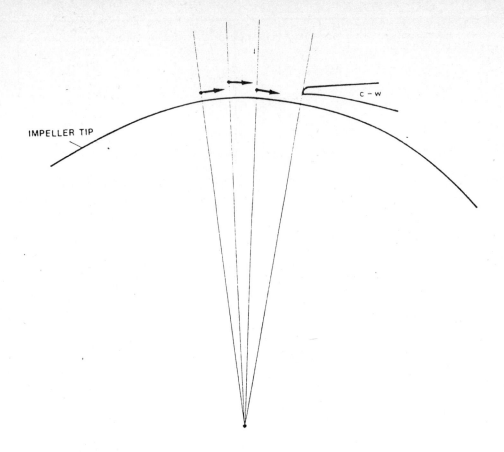

Fig 5a Absolute velocity vector near the cut water for 27 per cent BEP

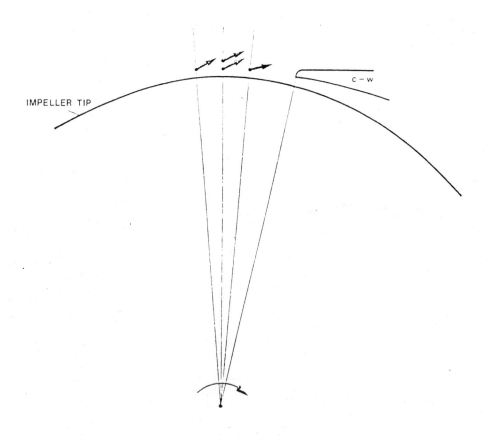

Fig 5b Absolute velocity vector near the cut water for 123 per cent BEP

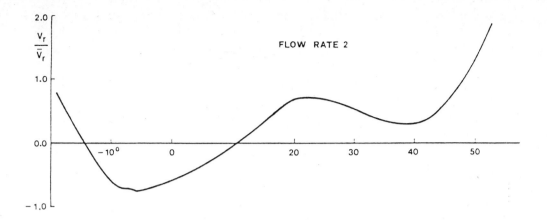

Fig 6a   Radial velocity distribution close to cut water for the 55.6 per cent BEP flow rate

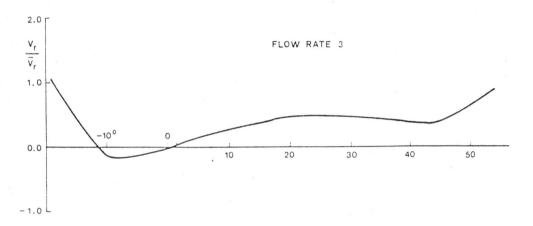

Fig 6b   Radial velocity distribution close to cut water for the 103 per cent BEP

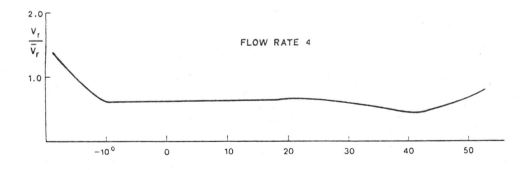

Fig 6c   Radial velocity distribution close to cut water for the 123 per cent BEP

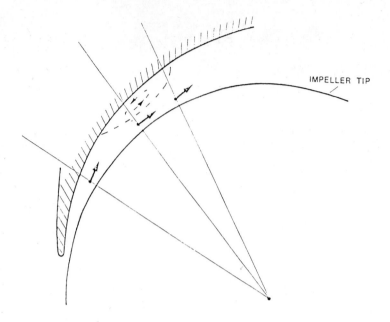

Fig 7   Velocity distribution downstream of the cut water

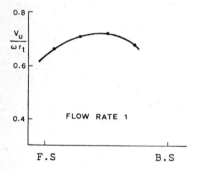

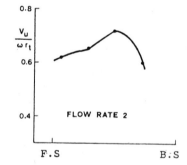

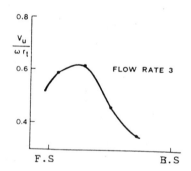

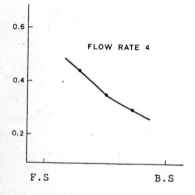

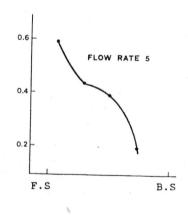

Fig 8

C337/88

Recirculation at impeller inlet and outlet of a centrifugal pump

G CAIGNAERT, J P BARRAND and **B DESMET**
Laboratoire de Mécanique de Lille, ENSAM, Lille, France

SYNOPSIS Previous measurements results reported on the "S.H.F. impeller" allow for a satisfactory determination of the recirculation critical flow rate at impeller inlet, but the methods used show a lack of precision in detecting the reverse flow onset at the impeller outlet. This paper presents new results from air tests demonstrating the importance of instantaneous pressures measurements within the impeller, in connection with inlet recirculation ; it then shows the value of hot-wires measurements at the impeller outlet in clarifying the dynamic phenomena associated with the matching of diffuser and impeller. Finally, it gives a brief description of a new experimental program that is now in progress in our laboratory for partial flows analysis.

1. INTRODUCTION

Much experimental work was already done on the so-called "S.H.F. impeller" by some laboratories in France, Switzerland and Italy, members of a work group of the "Société Hydrotechnique de France". The main results were reported and demonstrated that the recirculation onset at impeller inlet could be defined with a good precision by all the laboratories, in spite of the different casings, with various measurement or visualisation techniques just in front of the impeller inlet. Pressure measurements on the hub part of the impeller confirm these previous results and give some information about the stalled zone development.

Pressure or velocity evolutions at impeller outlet in the partial flow range proved much more difficult to interpret because of the unsteady and threedimensional behaviour of the flow in that region. It is evident that the flow at the impeller outlet is influenced jointly by the impeller and by the diffuser and volute geometries also that there may be some confusion between the reverse flow onset in the impeller and a diffuser stall extension up to the impeller. New results, obtained with a two hot wire probe, are presented and analysed as a confirmation of this point of view.

Finally we describe a new test facility which is intended to allow for a better description of recirculation at the impeller outlet.

2. TEST EQUIPMENT

The main characteristics of the impeller design are :

- outlet diameter : D_2 = 516.8 mm
- outlet width : b_2 = 38.7 mm
- inlet diameter : D_1 = 284.5 mm
- number of blades : Z = 7
- outlet blade angle : β_{2b} = 22°

- nominal flow rate : Q_N = 0,502 m^3/s at 2500 r/min.

An extensive description of the air test rig used in this work was reported in previous papers ((1) to (4)). Figure 1 shows the main dimensions of the vaneless diffuser and of the volute of constant cross section.

3. RECIRCULATION AT IMPELLER INLET

Recirculation at the impeller inlet of a centrifugal pump has been well described and analysed by many authors (see for instance references (5) and (6)), and some more or less accurate correlations also predict the recirculation critical flow rate (references (5), (7) and (8)). Tests in air and water with the "S.H.F. impeller" (references (1) to (4)) using various experimental techniques (see reference (9) for description and comparison) brought nearly constant values of the relative critical flow rate, Q_{K1}/Q_N, between 0.64 and 0.68, in spite of the different casings (vaneless or vaned diffuser, pump or pump-turbine casing) and of different leakage flow rates. From these results, it can be concluded that recirculation onset at the impeller inlet is essentially a characteristic of impeller inlet design, especially in the tip region.

In order to confirm these results and to gain more information about the recirculation development within the impeller, a piezoresistive pressure transducer was located on the hub part of the impeller, just in the middle of a blade passage (see radial position on figure 1). The pressure signal was transmitted to a F.F.T. analyser by using telemetry. Figure 2 (a) shows the flow rate dependance of the non-dimensional mean pressure for two speeds of rotation. It can be noticed that similitude is rather good between the two series of measurements. One observes also a steep change of the curve slope for a relative flow rate around 0.68. A simple model allows for a crude explanation of this behaviour; by applying BERNOULLI'S equation, without losses between the impeller inlet and the measuring point P, one obtains :

$$p - p_o = \rho \left(\frac{u^2}{2} - \frac{w^2}{2} \right) \qquad (1)$$

with :

p : pressure at the measuring point (Pa)
p_o: total pressure at impeller inlet (Pa)
    assumed to be constant .
u : peripheral velocity at the measuring point
    (m/s).
w : relative velocity at the measuring point
    (m/s).
ρ : density (kg/m^3).

In fact, one has to take losses into account in order to be closer to real pressure evolution at partial flow rates. The main losses, in that part of the impeller, are related to what is commonly called incidence losses. The leading edge of the rotor blades is designed to match the direction of the inlet velocity vector for what we call the nominal flow rate Q_N. When the pump is operating at partial flow rates, the inlet velocity vector is mismatched with the leading edge angle of the blades : this mismatch causes additional losses. Reference (12) gives the following expression of that incidence losses, Δp :

$$\Delta p = K \rho \left(w_{u1} - w_{u1}^{\circ} \right)^2 \qquad (2)$$

with :

K   : loss coefficient (adimensional)
w_{u1} : peripheral component of relative velocity
      vector at impeller inlet (m/s).
w_{u1}° = w_{u1} for adaptation.

Assuming adaptation and absence of prerotation at the nominal flow rate, consideration of the inlet velocity triangles gives :

$$w_{1u} - w_{1u}^{\circ} = u_1 \left(1 - \frac{c_{m1}}{c_{1N}} \right) - \frac{c_{m1}}{tg\alpha_1} \qquad (3)$$

with :

u_1 : peripheral velocity at point on the leading edge on the streamline passing through P.
c_{m1} : meridional component of the absolute velocity at inlet.
c_{1N} : absolute velocity at nominal flow rate.
α_1 : angle between the absolute velocity and the peripheral direction at inlet.

This expression shows that the head loss increases along the leading edge, from hub to shroud, and that, for any position, it increases as the flow rate is reduced as long as prerotation is not modified ($\alpha_1 = \alpha_{1N} = \P/2$).
Assuming now an uniform velocity distribution in the flow passages, an estimation of $p - p_o$ is then given by :

$$p - p_o = \rho \left(\frac{u^2}{2} - \frac{Q^2}{2S^2} \right) - K\rho \left(u_1 \left(1 - \frac{Q}{Q_N} \right) - \frac{Q}{S_1 tg\alpha_1} \right)^2 (4)$$

with :

S   : flow passage area at the measuring station.
S_1 : flow passage area at the blade leading edge.

This expression contains the whole information necessary to explain the main features of the evolution of pressure p at the measuring point, as schematically interpreted on figure 2 (b). During the first phase of the reduction (part (a) of the curve), S, S_1, and α_1 can be considered as constant. Pressure is then directly controlled by the decrease of Q, i.e. w, and by the increase of the head loss. It must be noticed that, owing

to the loss increase along the leading edge, secondary flows are induced which allow for the beginning of recirculation at the shroud for a relative flow rate of 0.68. This value appears as a slope discontinuity for the reason that, from this flow rate, the effective flow passage areas are decreasing.
Part (b) of the curve is governed by this new evolution. As it is well known however, some portion of the flow actually traversing the impeller is affected by some prerotation, the main effect of which being to reduce the mismatch of the incident velocity on the leading edge ($\alpha_1 < \P/2$), thus reducing the head loss. This progressive adaptation is noticeable on part (c) of the curve where the dimensionless mean pressure tends towards 1. At last, at very low flow rates, p_o must be also affected by the recirculation in the inlet pipe (part (d) of the curve). Experimental results (figure 2 (a)) correlate rather well with that rather crude model, which confirms that recirculation at inlet must be associated with stall and secondary flows in the inlet tip region of the blade and the shroud, with a rather large extension within the impeller. That extension could be studied with help of a larger number of pressure transducers in the impeller. These results suggest also that it would be possible to modify recirculation critical flow rate at inlet with a new design of blade leading edge, especially in the tip region.

Figure 3 shows the evolution, with the relative flow rate, of root mean square values of pressure fluctuations within the impeller at the impeller frequency of rotation (main component of frequency spectrum). The existence of these pressure fluctuations essentially comes from non-axisymetry of the flow within the casing ; however, they are greatly reduced as the pressure transducer is near the impeller inlet. Here again, it can be noticed that similitude is rather good between the measurements at two speeds of rotation, in spite of the low levels that are measured.
The increase of fluctuations between nominal flow rate and recirculation onset can be linked to static pressure evolution around the impeller outlet and to stall fluctuations on the suction side of the blades. The sharp decrease of fluctuations must be explained as a consequence of diminution of stall due to section area reduction and prerotation in the core of the flow. It must be noted also that we did not observe any occurence of frequencies non synchronous with impeller rotation in spectra, even at low relative flow rates ; this result indicates that we do not have rotating stall within this impeller as it is often found in centrifugal fans or compressors (references (10) and (11)).

4. RECIRCULATION AT IMPELLER OUTLET.

The value of the critical flow rate at the outlet of an impeller is generally given with a rather large uncertainty Some discrepancies were observed between the results initially obtained by the different laboratories engaged in the S.H.F. work group. From visualisations and measurements in air ((2), (3)) the onset of the reverse flow appears on the shroud side for relative flow rates between 0.72 and 0.9 according to the type of diffuser and to the speed of rotation of the impeller. Tests in water ((2) to (4)) confirm that the critical flow rate at outlet is greater than at inlet for the S.H.F. impeller. For instance, recent work made on the HYDROART test rig (reference (4)) brought an evaluation of the critical relative flow rate between 0.64 and 0.8 according to the interpretation of different measurements (axial and radial thrust fluctuations ; torque fluctuations ; pressure pulsations ; differential pressures between hub and shroud at diffuser inlet ; global characteristics).

There are several factors which explain the difficulty encountered in the precise evaluation of the critical flow rate ; the main reason for which is the unsteadiness of the flow coming out from the impeller :

a/ Visual observation of the very beginning of the reverse flow is difficult.

b/ The influence of the recirculation onset on the global mean characteristics (head or torque) is small and use of mean values of pressure or velocity can suffer from large discrepancies depending on the averaging technique.

c/ Fluctuating parameters are greatly influenced by the diffuser or the volute casing but also by the dynamic behaviour of the flow inside the whole test rig (especially when analysing pressure fluctuations in water tests).

In order to get a better interpretation of the different results obtained by the various laboratories involved in the study of the S.H.F. impeller, it seemed necessary to make instantaneous velocities measurements. First results of laser measurements in water (references (2) and (3)) and of hot-wires measurements in air (references (2) to (4)) show all the benefit that could be obtained with a good treatment of the results associated with these techniques. We shall now describe new measurements obtained with hot-wires in air to show how they can be correlated with more **traditional** measurements and interpretated to analyse recirculations.

Figure 1 shows the position of the probe that has been used for the velocity measurements: the hot wire support can be moved axially at a constant radius ($R/R_2 = 1.06$) and rotated around its axis. The two electrical signals coming from the two hot wire anemometers (DISA X-array probe 55P63) are sampled in a transient recorder (512 samples for one impeller revolution) ; data are transported to a micro-computer, stored, and transformed in radial and circumferencial velocity components according to the results of calibrations of each wire. The storage of each block of 512 samples is triggered by a pulse delivered by one signal synchronous to the impeller rotation in order to obtain synchronous averages. For that treatment, we assume that only the axial component of velocity is negligible. Results reported here show, for three different flow rates, the radial (fig.4) and circumferencial (fig.5) velocity components during one impeller revolution, at three different axial positions of the probe : near the shroud (3mm from the diffuser wall), in the middle of the diffuser width and near the hub (3 mm from the diffuser wall). For each flow rate and each axial position, one can see :

a/ the mean velocity component distribution during one impeller revolution (synchronous average over 8 revolutions).

b/ the superposition of the velocity component distributions corresponding to the 8 revolutions that were used for the average.

At design flow rate ($Q/Q_N = 1.0$), blade passing frequency appears well on velocity distributions in the middle and in the shroud part of the diffuser channel, and there is only a little dispersion from one revolution to another. The wake of each blade appears well on the radial component distributions and the circumferencial components have the rather conventional "sawtooth" form from pressure to suction side of each passage. Results are quite different near the hub : blade passing frequency does not appear well neither on radial nor on circumferencial components ; velocity distributions are more erratic, show a great dispersion from one revolution to an other one and even negative values of radial components appear from time to time. The same type of distributions, but with less dispersion has also been observed at 10 mm from the hub wall. So, even at impeller design flow rate, we notice a quite unsteady and non-periodic flow on the hub side of the diffuser inlet. This correlates well with a steady stalled zone on the diffuser wall as was demonstrated with mean velocity measurements made with a four-hole probe (reference (1)). In our opinion, we think that this stalled zone is not associated with recirculation in the impeller outlet but with stall on the diffuser wall as confirmed also by a theoretical model based on SENOO's assumptions (reference1).

At partial flow rates ($Q/Q_N = 0.742$ and 0.608 : figures 4 and 5), there is a shape alteration of the velocity distributions, mainly on the shroud side and in the middle of the diffuser width. Radial and circumferential components show a large scatter from one revolution to another, and we would have to make more than 8 recordings in order to get a stationnary synchronous average. Near the shroud this dispersion is more important and the periodicity at blade passing frequency disappears on both components ; radial components become negative from time to time but in a manner that cannot be only associated with blade wakes ; the negative parts of radial components are more and more important as flow rate is reduced. In the middle of the diffuser channel, in spite of the growing scattering of results as the flow rate goes down, one can always see the blade passing frequency especially on circumferencial components ; radial components become negative for only very short times not associated with blade passages. Near the hub, phenomena keep the same form as observed at impeller nominal flow rate ; it must only be remarked that negative values of radial components become more and more important as the flow is reduced. We see also that we can have a mean radial component positive with many instantaneous negative values and this demonstrates the necessity of instantaneous measurements in order to analyse with precision phenomena such as stalls or recirculations.

The comparison of these results with the mean velocities that we obtained with a four-hole probe (reference (1)) leads to the following attempts at interpretation :

a/ velocity distributions near the hub part of the diffuser must be associated with stall or separation on the whole diffuser wall, even at impeller nominal flow rate.

b/ At a relative flow rate equal to 0.742, velocity distributions near the shroud do not correlate with stall or separation on the diffuser wall. We think that the large dispersion of velocity from one revolution to another, with negative values of radial components appearing from time to time, reveals phenomena that occur within the impeller outlet and could be attributed to recirculations in that part of the impeller.

c/ At lower flow rate (Q/Q_N = 0.608), results are even more complicated because of the presence of stall or separation all along hub and shroud walls of the diffuser as was revealed by stationnary measurements (reference (1)).

5. CONCLUSIONS

The present paper shows important information that may be deduced firstly from pressure measurements within the impeller and secondly from instantaneous velocity analysis within the inlet of the diffuser.

An example of pressure measurement on the hub part of the S.H.F. impeller indicates how that type of information may be used to analyse the development of a stalled zone at the impeller inlet associated with inlet recirculation. The use of more transducers, on the hub or on the blades, would give a better description of basic phenomena that can explain the occurence of recirculation : such information is fundamental to the development of prediction methods.

The preceding analysis of velocity measurements explains why conventional methods whilst easier to use (for example, measurements of mean pressures or velocities) are not able to give full information about the reverse flow onset. On the contrary only instantaneous velocity measurements give a proper description of the unsteady and threedimensional features of the flow in this region, which allows for a separation between the diffuser or impeller disturbed behaviour at partial flow rates.

Partial flow behaviour study of the S.H.F. impeller is still in progress in our laboratory, taking into account previous results summarized in this paper. The impeller is now fitted with a new transparent volute. Conventional visualisations and measurements will be completed with instantaneous information. The impeller is now equipped with pressure transducers on suction and pressure sides of one blade, near outlet ; pressure signals on that blade will be correlated with pressure signals on the cutwater and with hotwires information in order to look at interaction phenomena between impeller and volute, especially at partial flows. This new experimental program will extend over 1988 and we hope that we shall have the opportunity of presenting some first results as a suplement to this paper.

REFERENCE

(1) CAIGNAERT, G., DESMET, B., MAROUFI, S., BARRAND, J.P.
Velocities and Pressures Measurements and Analysis at the Outlet of a Centrifugal Pump Impeller.
ASME Paper, 85-WA/FE-6, 7 pages.

(2) BARRAND, J.P., CAIGNAERT, G., CANAVELIS, R., GUITON, P.
Experimental Determination of the Reverse Flow Onset in a Centrifugal Impeller.
Proceedings of the 1st International Pump Symposium.
HOUSTON (U.S.A.) ; may 1984 ; pp. 63-71.

(3) BARRAND, J.P., CAIGNAERT, G., GRAESER, J.E., RIEUTORD, E.
Synthèse de résultats d'essais en air et en eau en vue de la détermination des débits critiques de recirculation à l'entrée et à la sortie de la roue d'une pompe centrifuge.
La Houille Blanche, 1985 n°5, pp. 405-419.

(4) LAZZARO, B.
Discussion of paper 45.
13 th IAHR Symposium ; section on hydraulic machinery, equipment and cavitation ;
Montreal (CANADA) ;
September 2-5 1986 ; session VIII B ; Flow in Pumps ; 5 pages.

(5) SEN, M., BREUGELMANS, F., SCHIAVELLO, B.
Reverse flow, prerotation and unsteady flow in centrifugal pumps.
Proceedings of NEL Fluid Mechanics Silver Jubilee Conference ;
Nov. 27-29, 1979 ; Paper 3.1. ; 30 pages.

(6) SCHIAVELLO, B.
Débit critique d'apparition de la recirculation à l'entrée des roues de pompes centrifuges : phénomènes déterminants, méthodes de détection, critères de prévision.
La Houille Blanche, 1982, n° 2-3, pp. 139-158.

(7) FRASER, W.H.
Recirculation in Centrifugal Pumps.
World Pumps, 1982, 188, pp. 227-235.

(8) PALGRAVE, R.
Operation of centrifugal Pumps at Partial Capacity.
9th. Technical Conference of the British Pump Manufacturer 's Association : Reliability - The User - Maker Partnership ; 16-18 April 1985 : Warwick University, Coventry ; Paper 6 ; pp. 57-70.

(9) CAIGNAERT, G., CANAVELIS, R.
Recensement et examen critique des méthodes expérimentales de détection des recirculations dans une pompe centrifuge.
13th IAHR Symposium ; Section on hydraulic machinery, equipment and cavitation ;
MONTREAL (CANADA) ; September 2-5 1986 ; Session VIII B : Flow in Pumps ; Paper 45 ; 15 pages.

(10) CAIGNAERT, G., DESMET, B., STEVENAERT, D.
Experimental Investigations on the flow in the Impeller of a Centrifugal Fan.
ASME Paper, 82-GT-37, 6 pages.

(11) FRIGNE, P., VAN DEN BRAEMBUSSCHE, R.
Distinction between Different Types of Impeller and Diffuser Rotating Stall in a Centrifugal Compressor with Vaneless Diffuser.
ASME Journal of Engineering for Gas Turbines and Power ; April 1984 ; vol : 106 ; pp. 468-474.

(12) BALJE, O.E.
Turbomachines : a guide to design, selection, and theory.
John Wiley & Sons - 1981.

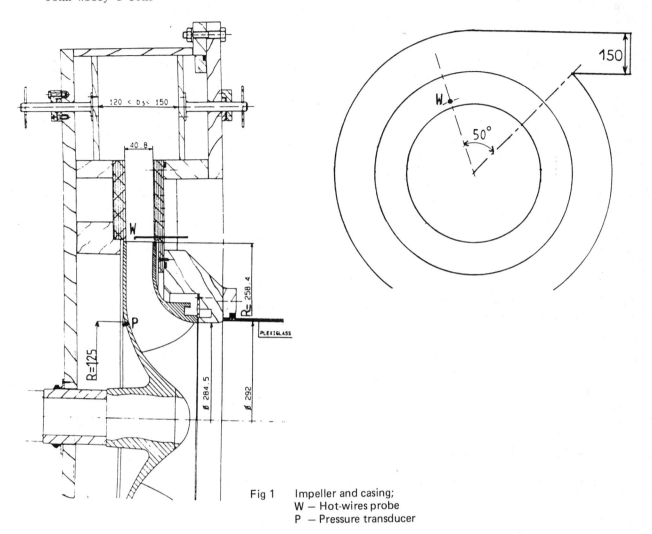

Fig 1   Impeller and casing;
W — Hot-wires probe
P — Pressure transducer

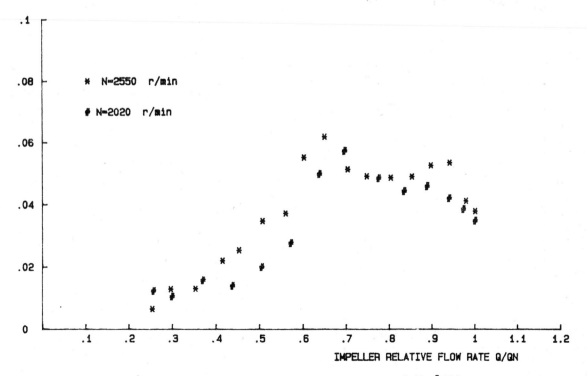

Fig 3   Dimensionless rms amplitude of pressure fluctuation $[p/(\rho u^2/2)]$
at impeller rotation frequency

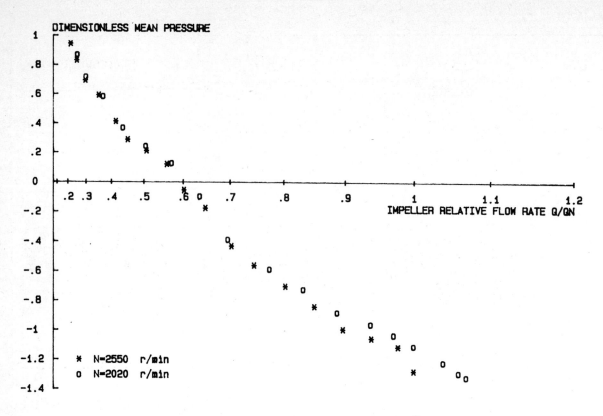

Fig 2a   Dimensionless mean pressure $(p\text{-}p_0)/(\rho u^2/2)$

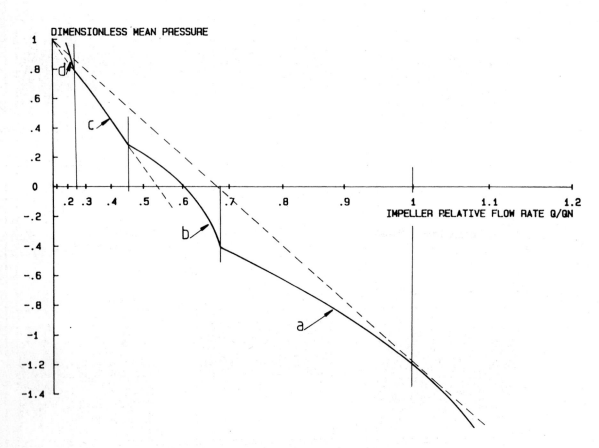

Fig 2b   Interpretation of dimensionless mean pressure evolution

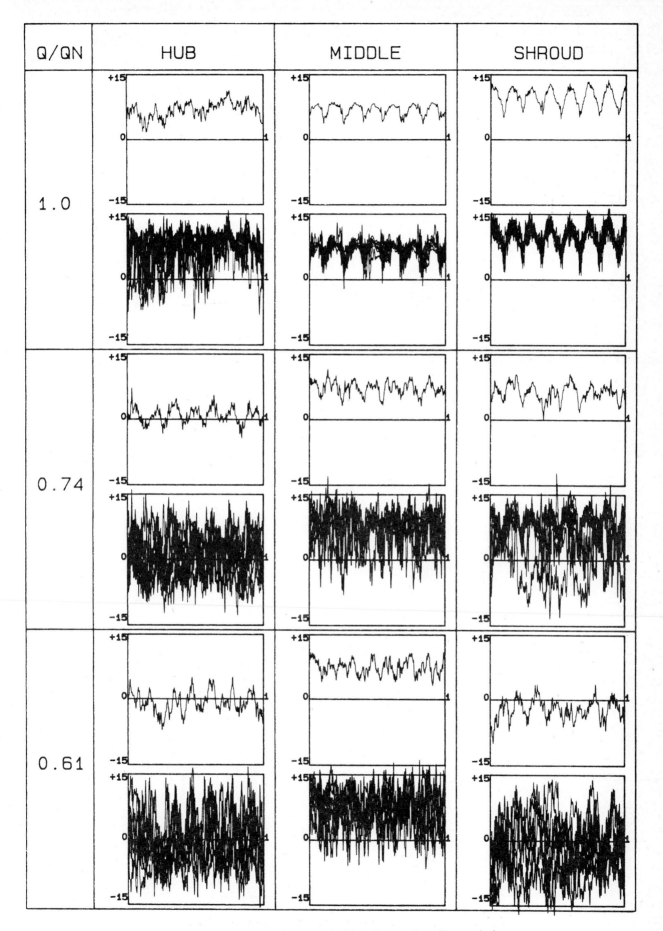

Fig 4   Velocity radial components (m/s) during one impeller revolution

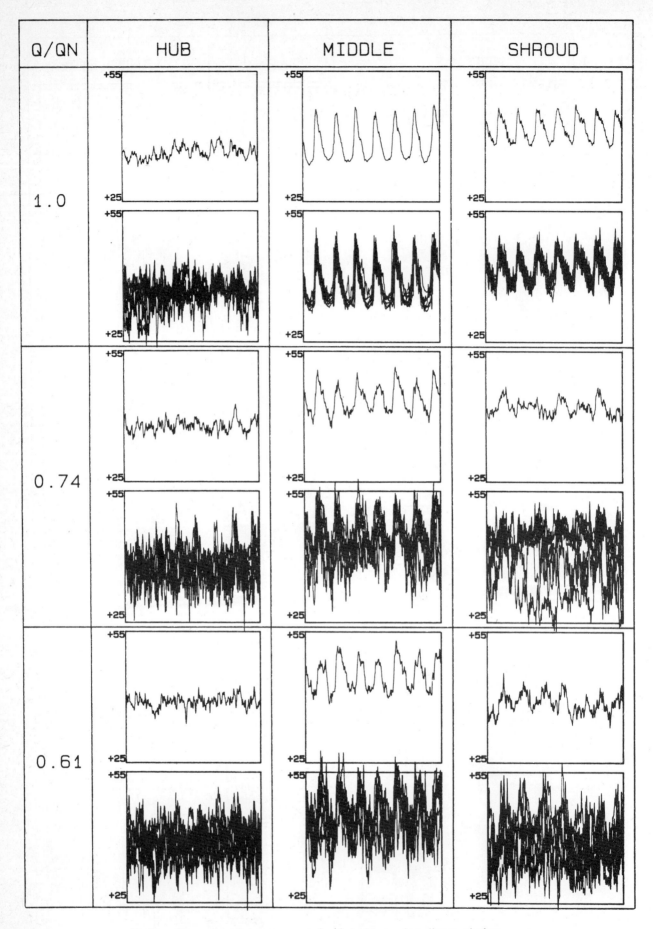

Fig 5   Circumferencial components (m/s) during one impeller revolution

C338/88

The flow field of two pump impellers operating at part-load

P HERGT, Dipl-Ing and **H JABERG**, Dr-Ing
Klein, Schanzlin und Becker AG, Frankenthal, West Germany

SYNOPSIS   Two pump impellers were investigated in air. The leading edge of one of them (impeller 2) was cut back compared to the other one (impeller 1). The onset of suction recirculation was considerably reduced by cutting back the leading edge, but an irregularity of the static pressure head curve was not influenced. For the original impeller 1 the flow rate of this irregularity coincides with the onset of suction recirculation, whereas for impeller 2 suction recirculation started at a much smaller flow rate. It is shown that the irregular form of the characteristic is influenced by a typical pattern of the discharge flow. The pressure side velocity distribution does not reflect the presence of suction recirculation.

1   INTRODUCTION

Part load pumping mostly is of crucial importance in the pump characteristics. Head curves may become unstable in different part load regions: - close to the optimum where we find the full load instability (1) - close to the onset of suction recirculation - close to zero flow with fully developed suction recirculation. KSB has long been performing measurements with impellers both in water and air to clarify this phenomenon of unstable characteristics. While trying to influence the onset of suction recirculation, an irregularity of the static pressure head curve was found. These measurements were performed in air where the static pressure head correlates closely to the total pressure head. The observed irregularity had the form of a dent in an otherwise continously increasing characteristic. The irregularity occurred either at the same flow rate as the suction recirculation or at a flow rate considerably larger.

It was found that the above irregularity may coincide with the onset of suction recirculation, although being obviously independent. The reason for this irregularity was a recirculation zone at the shroud on the impeller pressure side. The pump characteristics recovered when a thick wake also occured at the impeller hub wall. The wake position can closely be related to this head curve irregularity.

Since then the finding reported in this paper was confirmed by measurements with two other pumps having unstable characteristics.

The irregularity as well as the instability disappeared when the diffusers were removed, regardless whether vaned or not. Pressure side recirculations were much weaker or completely vanished.

2   EXPERIMENTAL SET-UP

A schematic drawing of the experimental set-up is shown in fig. 1a with a sectional view of the impellers investigated in Fig 1b. The air flow enters the pump through a suction pipe of 10.3 m length and 198 mm diameter, which is equipped at its entrance with a convergent inlet nozzle to measure the flow rate, an iris valve to adjust the flow and a stratifier to compensate uneven velocity distributions. The last 2287 mm upstream of the pump are equipped with 27 wall pressure taps, the first eight of which are located at a distance of 30 mm away from each other and the others at 100 mm.

The pump is driven by a V-belt drive with 2670 r/min. Both impellers' inlet diameter is 195 mm and the outlet diameter 395 mm. The outlet width is 24 mm in both cases. The leading edge angle is 21 ° at the shroud and designed for no-swirl flow. The trailing edge angle is 32 °. Both impellers differ in their meridional geometry (see Fig 1b): the leading edge of impeller 2 has been cut back and the shroud curvature is slightly smoother compared to impeller 1. Both impellers have seven blades. The flow rate of shock-free entry is $Q = 0.23$ m^3/s for impeller 1. As the leading edge of impeller 2 has been cut back without adjusting it, the flow rate of $Q = 0.23$ m^3/s will result in a shock-free entry only at the shroud, whereas the flow rates of shock-free entry at the inner flow paths will be slightly larger.

The pump is equipped with a vaneless diffuser of 27 mm width, 396 mm inlet diameter and 460 mm outlet diameter. Single hot-wire anemometry was applied to gain the results. The probe was located at $r = 203$ mm, i.e. 5.5 mm behind the impeller outlet.

The hot wire was 3 mm long and 5 μm in diameter. It was welded to two needles of 0.2 mm in diameter and 5 mm length. The support was 2 mm in diameter for the first 18 mm after which its diameter was 4 mm.

3 MEASUREMENT TECHNIQUE

The velocities were measured by hot-wire anemometry. The blade pressing frequency at 2670 r/min is 311.5 Hz. This frequency is well below the range of conventional hot-wire probes which are thus capable of nearly instantaneous measurements. To get a three-dimensional result the hot wire signal, i.e. its electric potential is recorded at up to 200 subsequent, separate intervals, each corresponding to a certain angle of revolution of the impeller. This measuring sequence is repeated a hundred times and the final signals at all the 200 angular positions are thus the arithmetic mean of hundred single recordings. As the hot-wire anemometer response is a function of the turning angle around the anemometer axis, the sequence described above must be repeated for different angles which will yield a sinusoidal signal distribution as a function of the anemometer angle for each of the 200 angular impeller positions. As is well known in hot wire anemometry, the flow angle is gained by fixing the minimum of the anemometer reponse, and the velocity at a setting 90 ° further. Once the absolute velocity and the absolute angle are known, all the other kinematic data can be calculated as the impeller tip speed is known.

The rotational speed of the pump is controlled by a digital speedometer when the test system is adjusted. During the measurements it is checked by the trigger frequency.

The hot-wire output is fed to a personal computer through a built-in A-D-converter, a mainframe was used only for plotting the results.

A three-dimensional distribution was assembled by measuring the angular velocity distributions at different depths, from hub (line A in Figs 3 and 4) to shroud (line G). As only the velocity component in the plane of the hot wire can be measured, no information can be gained concerning an axial velocity at the impeller outlet. A more comprehensive description of this measurement technique can be found in ref. 2.

4 STATIC PRESSURE HEAD CURVES

The static pressure head curves (Fig 2) are negative as the pressure at the outlet of the vaneless diffuser is the atmospheric pressure and the pressure at the suction side of the pump is decreased below its atmospheric value by acceleration and pressure losses mainly over the iris valve.

The head curve of impeller 1 (Fig 2a) shows an irregularity right at the onset of suction recirculation. When the flow rate is decreased the head curve changes form steeply increasing to horizontal and increases sharply after the onset of recirculation. Then the head curve continues to increase with a gradient as before until the suction recirculation extends to the location of the wall pressure tap.

When no recirculation is present the pressures at different upstream locations coincide as the short distance between them hardly causes a pressure drop. But when suction recirculation has formed, the strongly rotational flow exerts a radial pressure gradient that approximately balances the pressure drop in the throttling device.

Thus the extent of the suction recirculation is clearly visible.

The head curve irregularity, which may not necessarily result in an unstable pump characteristic, was believed to be caused by the onset of suction recirculation, the details of which are not known. Obviously the losses increased inside the impeller as no reason could be found for extra losses in the suction side. When the recirculation zone extended into the suction pipe, the pressure recovered. We expect evidence of these flow conditions from measurements in the rotating system which have not been performed so far.

When the static pressure head curve of impeller 2 (Fig 2b) was measured, we found that suction recirculation set on at a much smaller delivery showing the same phenomena as described before. But the head curve now had a second irregularity at a slightly larger flow rate than the curve of impeller 1. Measurements of the flow at the suction side of both impellers showed no difference between them (3). Thus it was concluded that the irregularity of impeller 2 was caused by some alterations of the flow conditions inside or at the outlet of the impeller which acted through the impeller until into the suction pipe. If this was true then impeller 1 should suffer from the same circumstances, with the only difference that it coincides with the suction recirculation onset.

It is worth mentioning that the onset of suction recirculation seems to depend only on the impeller inlet geometry, but is not influenced even by strong modifications of the outlet geometry (4).

5 MEASUREMENT OF PRESSURE SIDE VELOCITY DISTRIBUTIONS

To find the reason for the above mentioned form of the pump performance, detailed measurements of the velocity distribution at the impeller pressure side were performed. The pitchwise radial velocity distribution is shown in Fig 3 and the tangential velocity in Fig 4, both of them for different flow rates. Line A always refers to a location 2 mm away from the shroud and line G 2 mm away from the hub. Line D represents the pitchwise velocity distribution approximately at the diffuser centreline, i.e. 14 mm away from the shroud. The position of the impeller blade is indicated by S and P for suction and pressure faces of the vane.

5.1 Radial velocity

The results for the radial velocity are shown in Fig 3a and Fig 3b for impeller 1 and impeller 2, respectively. At a flow rate of $Q = 0.25$ m^3/s at b.e.p., the velocity distribution is more or less symmetric over most of the pitch. On the suction side of the vane the wake region can be seen as a region of low velocity which is especially pronounced at the shroud (line A). Between the diffuser centre line and a thin wake layer on the hub hardly any wake can be detected.

When the flow rate is reduced, the radial velocity on the shroud continuously decreases for both impellers and finally becomes negative. At flow rates right before the recovery of the static pressure head curve from the dentlike irregularity, at $Q = 0.16$ m^3/s for impeller 1 and

$Q = 0.18$ m^3/s for impeller 2, the low radial velocity at the shroud extends over the entire pitch. For impeller 2 it is clearly negative whereas for impeller 1 it varies around zero, but the overall shape is very much the same. Close to the hub, the velocity (line G) for both impellers is somewhat reduced compared to the larger delivery in the middle of the blade passage which may be due to the wall boundary layer. For both impellers the peaks of the radial velocity are considerably lower at the shroud disk than on the hub disk whereas at $Q = 0.25$ m^3/s the velocities from hub to shroud for both impellers reached the same maximum values.

When reducing the flow rate below $Q = 0.25$ m^3/s the velocity peak in the depths E & F (between centreline and hub) vanished and the flow pattern of both diffusers show the existence of a dip or wake which is located at the blade suction surface and extends into the middle of the blade passage.

When further reducing the flow rate to a value where the static pressure head curve has recovered ($Q = 0.14$ m^3/s for impeller 1 and $Q = 0.16$ for impeller 2), the radial velocity at the hub is reduced to a value as low as the radial velocity at the shroud. This blockage effect is not confined to a narrow wall layer but extends deep into the diffuser passage (line E & F). For impeller 2 this hub layer 'fills' the backflow area that was present on the shroud at $Q = 0.18$ m^3/s.

Although this forming of a hub wake is more pronounced for impeller 2, because the hub velocity decreased from a higher level to zero, it is detectable in the impeller 1 flow field as well. Both impellers now have radial velocity distributions which are roughly symmetric over the diffuser passage. The velocity in the middle (line D) has remained unaltered when throttling from $Q = 0.18$ m^3/s to 0.16 m^3/s or $Q = 0.16$ m²/s to 0.14 m^3/s respectively.

When the flow rate is further decreased the suction vortex of impeller 1 starts to develop (see Fig 2a) and grows into the suction pipe. The slope of the static head curve is constant as long as the suction recirculation has not reached the upstream position of the wall pressure tap. The slope of the static head curve of impeller 2 (c.f. Fig 2b) is constant for all suction pipe pressure taps because the recirculation starts at $Q = 0.1$ m^3/s only.

When investigating the pressure side flow field at deliveries lower than the recovery rate, the flow patterns of both impellers are almost identical, regardless of the presence of the impeller 1 eye vortex. In the fourth part of Figs 3a and 3b, the radial velocities of both impellers at $Q = 0.1$ m^3/s are illustrated. At the hub and at the shroud large separation areas with strong backflow have developed and a distinct wake is detected that fills the entire blade passage. Around the centreline (letter D), however, the radial velocity has increased, in spite of reducing the delivery. So the lower flow rate has been verified at the expense of backflow along both discs. The flow patterns are symmetrical to the diffuser centreline.

The negative radial velocities at $Q = 0.1$ m^3/s have developed continuously when throttling below $Q = 0.14$ m^3/s (impeller 1) or $Q = 0.16$ m^3/s (impeller 2) though the interim results are not shown. Smaller flow rates basically show the same flow patterns as $Q = 0.1$ m^3/s. Remarkably, the discharge flow field shows no influence of suction recirculation at any delivery.

5.2 Absolute Tangential Velocity

The tangential velocities of the discharge flow are shown in Fig 4a for impeller 1 and in Fig 4b for impeller 2. The graphs correlate closely to those of the radial velocity Fig 3.

At the flow rate $Q = 0.25$ m^3/s both impellers have a dip behind the impeller vane and a sharp increase on the suction face of the impeller blade. Contrary to the radial velocities this increase is roughly equally spread over the diffuser width which means that in the wake at the shroud with low radial velocity the relative flow leaves the impeller quite tangentially and it steepens as the plane of measurements approaches the hub. Over most of the impeller quite a good energy transfer into the flow is found for both impellers as the radial and the tangential velocities are high.

When the flow rate is reduced, both impellers' radial velocity distributions become very oblique. At a delivery slightly larger than at static pressure head recovery, i.e. $Q = 0.16$ m^3/s for impeller 1 and $Q = 0.18$ m^3/s for impeller 2, the tangential velocity at the shroud is reduced to a value being just a little larger than in the wake behind the blading. The tangential velocity increases with the distance from the shroud and a peak develops near the blade suction face: for impeller 1 the peak reaches 40 m/s and for impeller 2 37 m/s, while the velocity outside the dip and peak is a little less than 36 m/s for both impellers. Very close to the hub (line G) the tangential velocity of both impellers is lower compared to the neighbouring wall distances. This is analogous to the radial velocity pattern and seems to be due to the wall boundary layer. At these two flow rates, most of the hydraulic work is obviously done by the hub of the flow passage.

When throttling the flow rate to its value below head curve recovery (impeller 1: $Q = 0.14$ m^3/s; impeller 2: $Q = 0.16$ m^3/s), the velocities at the hub (lines G and F) are suddenly seriously reduced, whereas the velocities at the shroud are increased. This finding is more pronounced for impeller 2 with the leading edge cut back: the velocity at the shroud (line A) is increased by 4 m/s and even the centreline tangential velocity is increased by 2 m/s. For impeller 1, the shroud side velocities from line A to D increase by 1 to 2 m/s. As it was found before for the radial velocities, the tangential velocities now are approximately symmetrical to the diffuser centreline, where most of the hydraulic work is done.

For lower flow rates the tangential velocity distribution of both runners remains symmetric and similar to the radial velocity pattern. No influence of the part load vortex is detectable. The results for impeller 1 (Fig 4a) with a strong suction vortex and impeller 2 (Fig 4b) with no suction vortex at this flow rate are almost identical. The fourth part of Figs 4a and b shows the tangential velocity profiles at Q = 0.1 m³/s. Only little has changed compared to the profiles at the next larger flow rate. The tangential velocity at the hub has slightly decreased whereas it has increased at the hub and along the centreline (line D)

In accordance with potential theory, the tangential velocity of both impellers along the centreline is higher at the part load delivery of Q = 0.1 m³/s than at b.e.p. with Q = 0.25 m³/s. But along the hub and shroud the radial as well as the tangential velocities are low at part load pumping.

5.3 Mean velocity variations

The instantaneous velocity distributions shown in Figs 3 and 4 and discussed in the previous paragraphs are circumferentially averaged and plotted in Fig 5 as a function of the flow rate for three axial stations: near hub, mid and near shroud. These graphs give evidence how the velocity profiles develop from b.e.p. to part load.

When the flow rate is reduced below b.e.p. but is still close to it, the tangential velocity increases and the radial velocity decreases. So at large flow rates the measurements agree well with potential theory although the shroud side is loaded less than the rest of the flow regime. The measured gradient of the radial velocity reduction coincides with the theoretical one for equally distributed radial velocity, it is shown in Figs 5a and b as a dashed line.

At the shroud the mean values of the radial velocity decreases faster than the mid and hub values. This process is paralleled by a decrease of the shroud side tangential velocity when the flow rate is further away from b.e.p.. At a certain flow rate, i.e. Q ≈ 0.15 m³/s for impeller 1 and Q ≈ 0.17 m³/s for impeller 2, both the tangential and the radial velocities at the hub begin to decrease steeply, the tangential velocity at the shroud starts to increase again, in accordance with potential theory, and the radial velocity increases, though only temporarily. This hub side breakdown even led to a slight increase of radial velocity in the middle of the diffuser and a faint decrease of the tangential velocity was brought to a stop.

When further reducing the delivery the tangential velocities increase and the radial ones decrease at all locations but at hub and shroud on a much lower level than at mid channel.

From Fig 5 it follows that a decrease in both the radial and tangential velocity must result in a very flat discharge velocity triangle, i.e. the relative velocity points almost in the opposite direction of the impeller tip speed or in other words, the impeller must have developed a large suction side wake. Fig 5 could indicate that this is only true at the hub and shroud and thus could possibly be caused by the wall boundary layer. But as the instantaneous measurements showed (see Figs 3 and 4) the pitchwise local velocity distribution may differ from the mean values because the wake may be balanced by the jet region resulting in high mean values. Vice versa it is concluded that the decrease of the mean values of the velocities (Fig 5) is not necessarily caused by an increase in wake size but rather by a movement of the wake.

5.4 Contours of radial velocities

Contour plots of the radial velocity over the cross section of the diffuser 5.5 mm behind the impeller exit are shown in Figs 6a and 6b for impeller 1 and impeller 2, respectively. In all parts of the picture the shaded area represents the wake region. The wake region is defined as the region with radial velocities lower than the average velocity in the plane of measurement. Backflow areas are shaded crosswise.

At Q = 0.25 m³/s a large wake region of both impellers can be seen which has its core in the shroud disk/suction face corner of the impeller vane some 15 - 20 p.c. of the blade pitch away from the suction face. It extends deep into the passage. In the hub disk/suction face corner a small jet-like region with high radial velocity is located that will disappear with throttling. As the plane of measurements is 4 p.c. of the impeller radius downstream the impeller exit, the jet has reached this position within this distance.

The next smaller flow rates shown do not indicate a distinct wake core, but both impellers now have a wake distribution that spreads out over all the pitch on the shroud disk and extends, where it is thickest, almost into the middle of the diffuser passage. The hub side wake region of impeller 1 is only very thin and is believed to be rather the normal boundary layer, impeller 2 shows nothing of this kind on the hub.

When reducing the flow rate to its value where the head curves have recovered, distinct wake regions develop on the hub and consequently the wake regions at the shroud become thinner, thus symmetrizing the flow pattern. At this flow rate the jet which was present at Q = 0.25 m³/s in the hub/suction face corner has completely vanished and only the usual jet is visible. When the second and the third distribution shown in Figs 6a and 6b are compared to each other it becomes clear that the flow has become symmetric. So the lines of constant velocity of the jet are half circles at Q = 0:16 m³/s (Fig 6a) and Q = 0.18 m³/s (Fig 6b) but closed circles at the next smaller flow rates. A weak backflow is present at both disks of impeller 1, but the impeller 2 backflow area which was present at Q = 0.18 m³/s has disappeared, this is taken as a consequence of the now symmetric radial velocity distribution.

When the delivery is throttled below the head curve recovery value, the radial velocity patterns remain symmetric from hub to shroud. Close to both disks distinct backflow areas have developed but the radial velocity level over most of the blade pitch has not changed. The lower delivery has mainly been verified firstly by a lower peak velocity in the jet and secondly by the backflow areas. A wake core can not be found at these low rates as we find thick layers of equally low velocities on both end disks.

6 DISCUSSION

Pressure side recirculations have often been reported in the literature. One of the earliest ones is the work by Schrader (5). Since then connections between either suction or discharge recirculations and the form of the pump characteristics have frequently been proved. Rey et al. (6) have shown that the critical flow rates for either of these recirculation onsets are independent and, although different, in general close to each other. They give, among others, graphs of the critical delivery as functions of the specific speed and the head coefficient. Detailed results concerning the flow patterns on the suction and discharge side of an impeller at the critical flow rates both in water and air test can be taken from the paper by Barrand et al. (7). Their instantaneous measurements showed a similar development of the onset of discharge recirculating flow although it was not correlated to a certain form of the characteristics. They found that suction separation always starts at lower flow rate than discharge separation.

The influence of a distortion of the impeller discharge flow on diffuser separation was studied in several papers by Senoo and Kinoshita (8, 9, 10). In ref. (9) they give a design rule, illustrated in several graphs, how the diffuser inlet angle in a possibly distorted flow can meet the requirements of avoiding separation. When applying these results to the impeller reported here, it turns out that the absolute flow angle at the diffuser inlet is in the range where recirculation is likely. This agrees with the results by Tsurusaki et. al. (11) who also find diffuser inlet angles at onset of recirculation that are smaller than predicted by theory.

Our findings are that at larger flow rates, $Q = 0.25$ m^3/s, the absolute angle at the shroud disk is well below its critical value as given by Senoo (9), although at this flow rate no recirculation occures but at smaller flow rates. So it is not sure whether in our case the shroud flow was stabilized, in spite of the small inlet angle, by a transport of energy from mid-channel to the shroud, or whether in the set-up investigated here the flow was so stable that separation occured at smaller flow rates only. Senoo and Kinoshita (8) agree with our results in so far as close to b.e.p. the radial velocity at the shroud is lower than at the hub and at part load it is higher than at the hub, but their measurements differ from ours in two respects, firstly the tangential velocity is roughly symmetric over the diffuser width and secondly no recirculation occures in their results at the diffuser inlet plane. So the recirculation on the shroud and at smaller flow rates also on the hub, will probably be caused by an appropriate wake development in the impeller and thus the diffuser flow is dominated by the impeller discharge flow. Similar results have been gained by Inoue (12) who found radial contours that agree well with the ones outlined in Fig 6.

The forming and the position of the wake have often been described, e.g. (13, 14). Lennemann and Howard (13) demonstrate the extent of the wake region for different flow rates by means of the hydrogen bubble technique. In their experiments they find a connection between discharge and suction recirculation because both separations are commonly caused by a suction face wake separation. From our experiments no evidence concerning this connection can be expected as only discharge and suction side measurements were performed. But as far as the discharge flow patterns are concerned, it became clear that they do not depend on the presence of an impeller eye vortex. Vice versa it has been outlined by Riegger (3) and Hartmann (4) that suction recirculation depends only on the inlet geometry. According to Johnson and Moore (14), the wake position depends on whether the flow is dominated by rotation or by curvature, a measure being the Rossby number. Our results seem to agree with refs. (12, 14) as at the higher flow rates the wake is situated only at the shroud, and at lower flow rates at the shroud and at the hub. The hub disk wake most probably originated from the shroud wake, because when the flow rate is diminished, curvature effects gain importance and the wake moves to the blade suction face, due to inertia effects it runs beyond its stable location and reaches the hub disk.

The separation region at the shroud that becomes thinner when the hub flow also separates has been measured before by Caignaert et al. (15). Their results also show a symmetric radial velocity distribution with recirculation along both sides at part load, but they did not give the pump characteristic. Theses results also agree with those given by Rebernik (16).

What is called wake in our measurements differs from the usual definition in so far as in the wake and recirculation areas our measurements showed that the relative velocity at the impeller outlet points approximately in the opposite direction of the impeller tip speed causing a tangential absolute velocity of roughly half the tip speed. Normally a wake is a region of low relative velocity thus causing a high tangential component of the absolute velocity. So the wake flow in our measurements tends to separation.

From the measurements described above it is concluded that the reason for the head curve irregularities is the wake that is present in the shroud disk/suction face corner and grows along the shroud disk. Parallel to the decreasing of the radial velocity at the shroud, the head curve changes from steeply increasing with decreasing flow rate to horizontal. When the wake develops at the hub, the flow becomes symmetric over the diffuser passage, the head curve recovers suddenly and the gradient of the characteristic gains the value outside this irregularity. This phenomenon is especially clear for impeller 2, but is detectable as well behind impeller 1. A possible cause for the differences is that by cutting back the leading edge of impeller 2, the blade length at the hub is reduced considerably which means that the blade loading increased as did the pressure gradient along the blade. Consequently the impeller 2 flow along the hub separated earlier than the impeller 1 flow. Besides increasing the blade inlet angle by cutting back the leading edge, this increases the shock of the flow and makes separation more likely. Nevertheless the flow rates where the static pressure head curves exhibited these irregularities are quite close to each other. As it is shown above, the tangential velocity decreases together with the radial velocity which means that the relative flow leaves the impeller almost tangentially, so that the cause of the irregularity can be found to some extent in the impeller. Roughly speaking, the diffuser is offered a flow condition it does not accept.

That the onset of suction recirculation coincides with the irregularity is obviously chance as the impeller 2 eye vortex starts at a much lower delivery but the discharge flow pattern and the form of the pump characteristic around the larger flow rate remain unchanged.

A flow pattern similar to that described above can be found in pumps having vaned diffusers and unstable characteristics. The forming of the instability is accompanied by an increasing wake at the shroud disk and the head curve recovers when the velocity profiles become symmetric.

Future work will have to concentrate on the point of finding design rules that allow for impellers with unconditionally stable characteristics. A possible way is to design the impeller in a way that controlls the wake formation of the blade suction face and always avoids suction face separation.

REFERENCES

(1) P. Hergt, J. Starke. Flow patterns causing instabilities in the performance curves of centrifugal pumps with vaned diffusers. 2 nd Int. Pump Symp., Houston 1985.

(2) C.P. Hamkins. Das "Periodic Sampling" Geschwindigkeitsmeßverfahren. KSB-report (unpublished), 1986.

(3) R. Riegger. Experimentelle Untersuchungen des TLW-Verhaltens verschiedener Varianten eines frei abströmenden Radialpumpenlaufrades mit axialer Saugleitung. Diploma thesis, KSB Frankenthal and FH Konstanz, 1985.

(4) O. Hartmann. Experimentelle Untersuchungen der Strömungsvorgänge am Eintritt eines frei abströmenden Kreiselpumpenlaufrades mit axialer Saugleitung. Diploma thesis, KSB Frankenthal and TU Karlsruhe, 1986.

(5) H. Schrader, Messungen an Leitschaufeln von Kreiselpumpen. Würzburg: Triltsch, 1939.

(6) R. Rey, Y. Keramarec, P. Guiton, G. Vullioud. Etudes statistiques sur les charactéristiques à debit partiel des pompes centrifuges et sur la détermination approximative du débit critique. La Houille Blanche, No. 2/3, 1982.

(7) J.P. Barrand, G. Caignaert, R. Canavelis, P. Guiton. Experimental determination of the reverse flow onset in a centrifugal impeller. 1st Int. Pump Symp., Houston, 1984.

(8) Y. Senoo, Y. Kinoshita, M. Ishida. Asymmetric flow in vaneless diffusers of centrifugal blowers. Gas Turbine and Fluids Engineering Conf., New Orleans, 1976.

(9) Y. Senoo, Y. Kinoshita. Influence of inlet flow conditions and geometries of centrifugal vaneless diffusers on critical flow angle for reverse flow. Gas Turbine and Fluids Engineering Conf., New Orleans, 1976.

(10) Y. Senoo, Y. Kinoshita. Limits of rotating stall and stall in vaneless diffuser of centrifugal compressors. ASME paper 78-GT-19, 1978.

(11) H. Tsurusaki, K. Imaichi, R. Miyake. A study on the rotating stall in vaneless diffusers of centrifugal fans (1st report, rotational speeds of stall cells, critical inlet flow angle). JSME International Journal, 1987, 30, 279-287.

(12) M. Inoue. Centrifugal compressor diffuser studies. Ph. D. Thesis, 1980, Department of Engineering, Cambridge University. M. Inoue, N.A. Cumpsty. Experimental study of centrifugal impeller discharge flow in vaneless and vaned diffusers. Journal of Engineering for Gas Turbines and Power, 1984, 106, 455-467.

(13) E. Lennemann, J.H.G. Howard. Unsteady flow phenomena in rotating centrifugal impeller passages. Journal of Engineering for Power, 1970, 92, 65-72.

(14) M.W. Johnson, J. Moore. The development of wake flow in a centrifugal impeller. Journal of Engineering for Power, 1980, 102, 382-389.

(15) G. Caignaert, B. Desmet, S. Maroufi, J.P. Barrand. Velocity and pressure measurements and analysis at the outlet of a centrifugal pump impeller. ASME paper 85-WA/FE-6, 1985.

(16) B. Rebernik. Investigation of induced vorticity in vaneless diffusers of radial flow pumps. 4th Conf. on Fluid Machinery, Budapest, 1972.

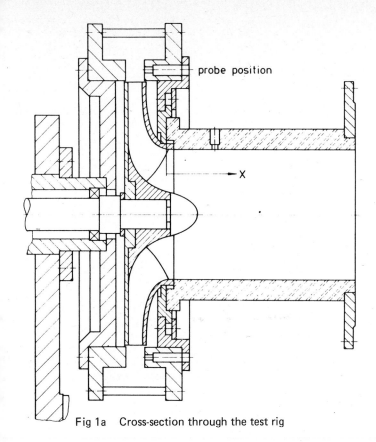

Fig 1a   Cross-section through the test rig

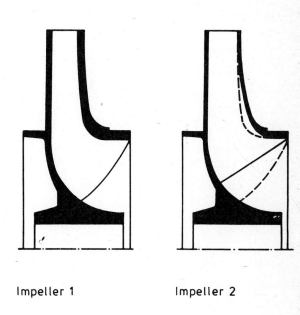

Impeller 1   Impeller 2

Fig 1b   Sectional view of the impeller investigated

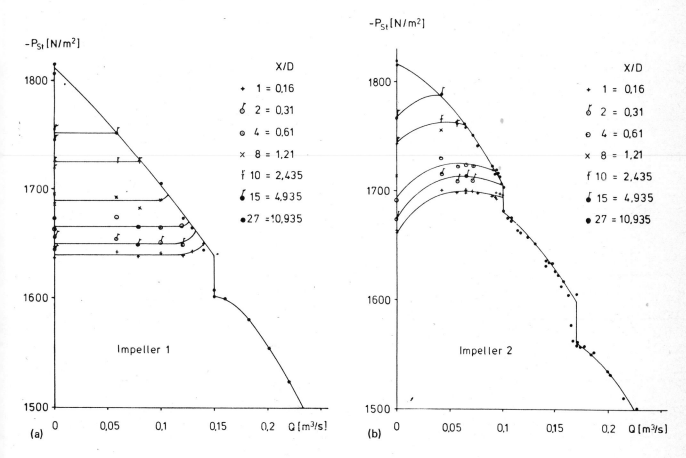

Fig 2   Static pressure head curves
(a) Impeller 1
(b) Impeller 2

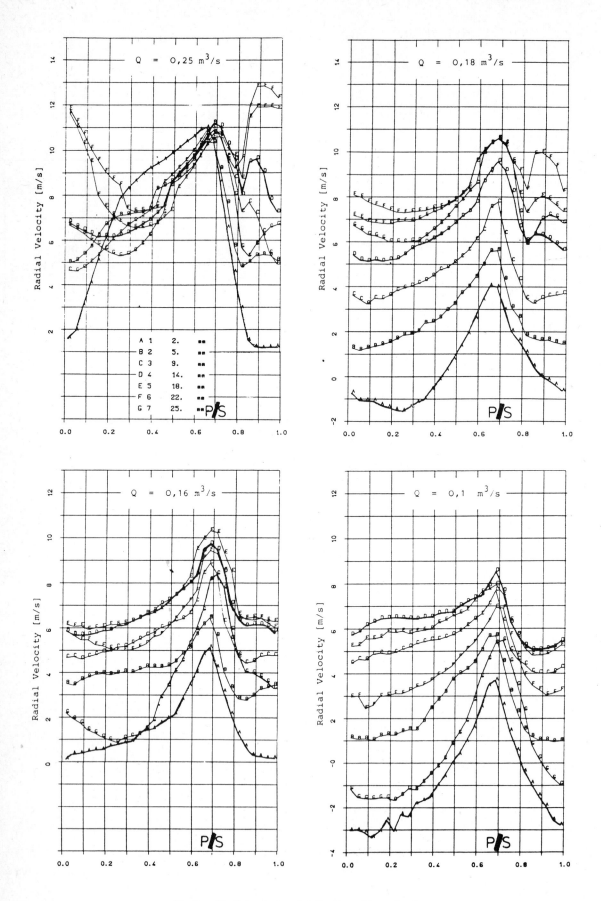

Fig 3a   Radial velocity distribution of impeller 1

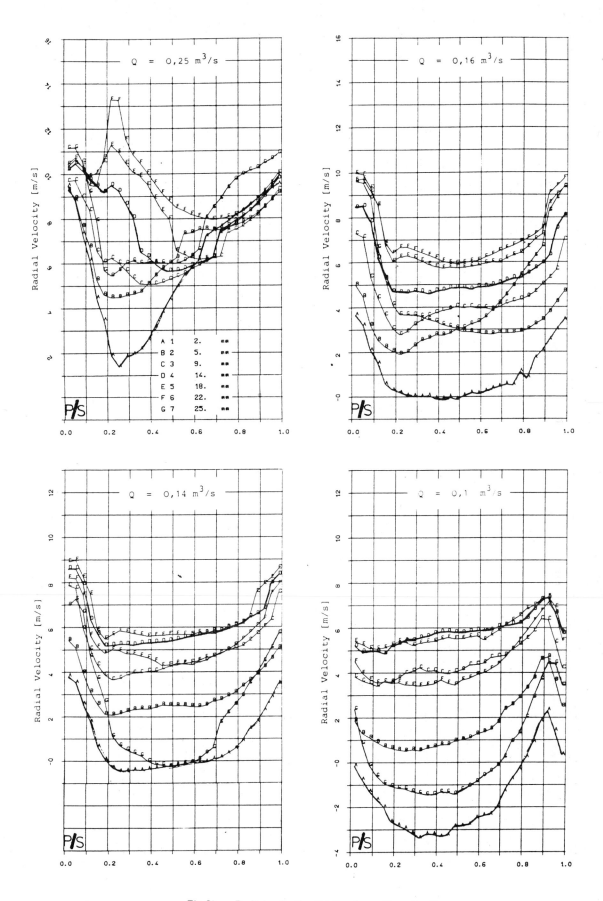

Fig 3b   Radial velocity distribution of impeller 2

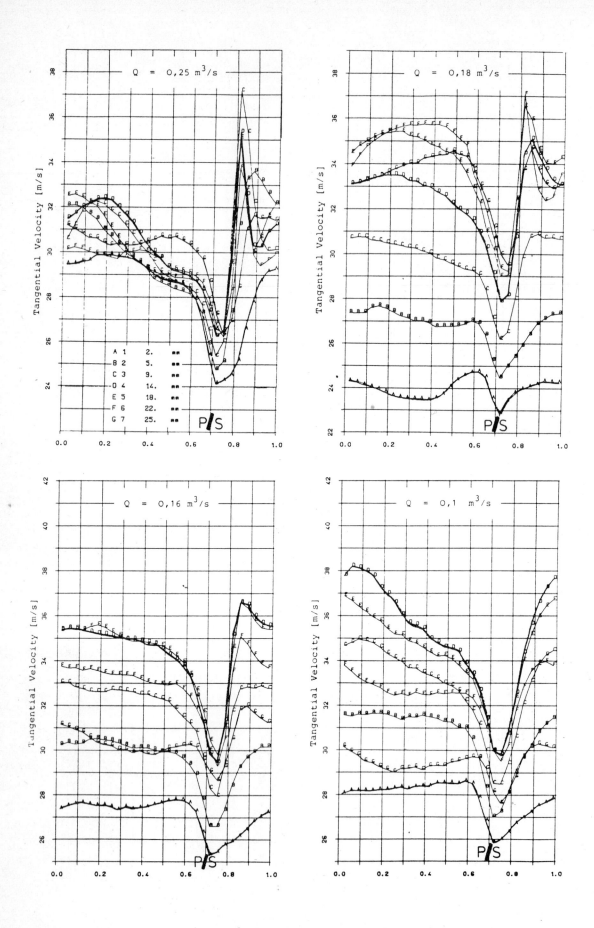

Fig 4a   Tangential velocity distribution of impeller 1

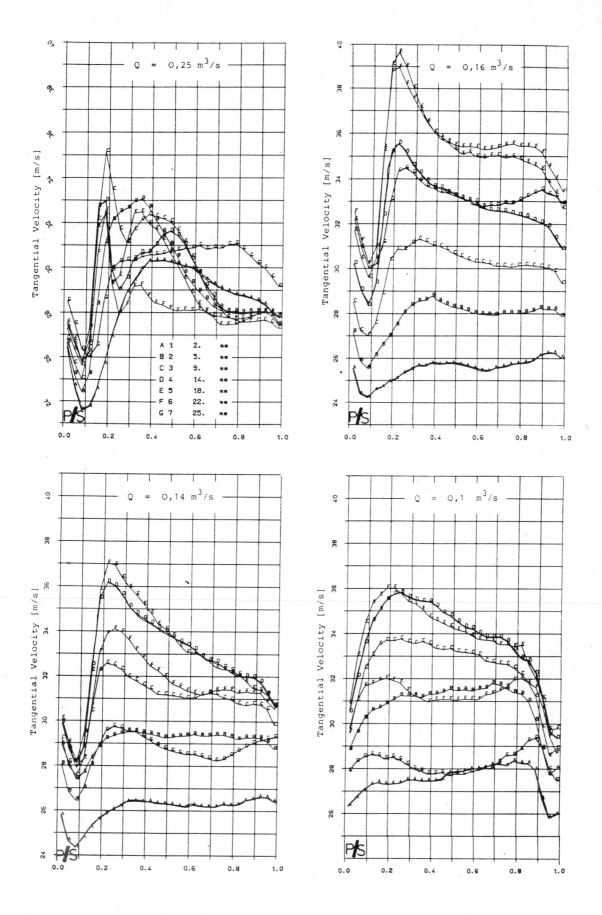

Fig 4b   Tangential velocity distribution of impeller 2

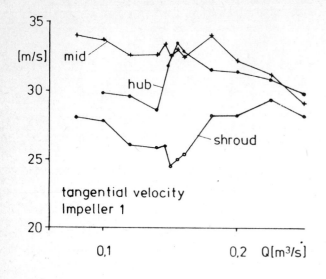

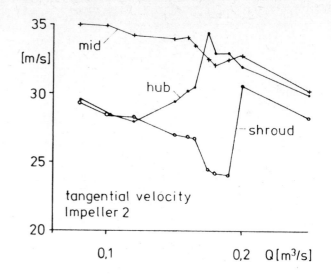

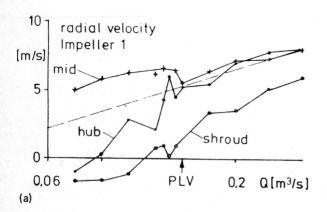

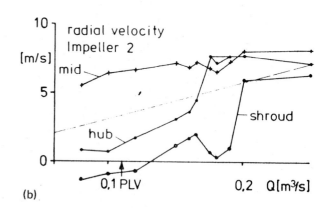

Fig 5   Circumferential averaged tangential and radial velocities as
function of the flow rate
(a) Impeller 1
(b) Impeller 2

Rotation

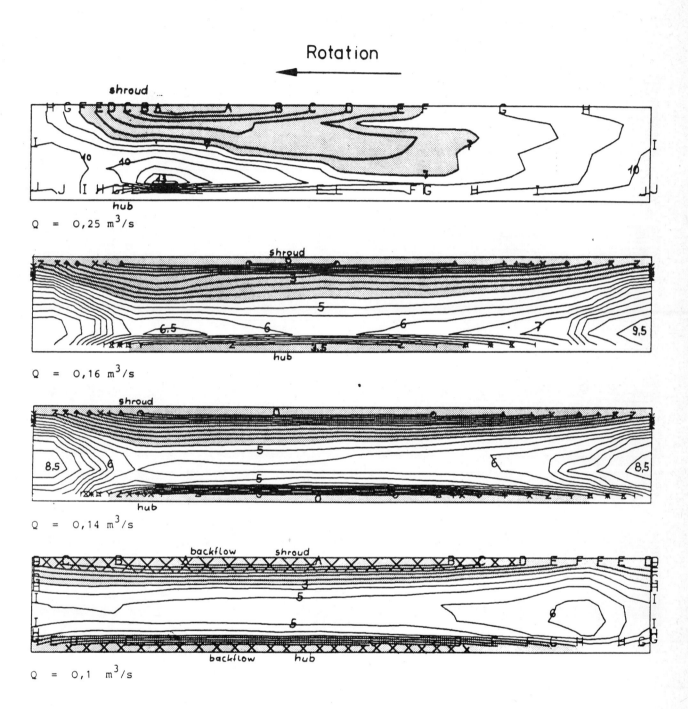

$Q = 0,25 \text{ m}^3/\text{s}$

$Q = 0,16 \text{ m}^3/\text{s}$

$Q = 0,14 \text{ m}^3/\text{s}$

$Q = 0,1 \text{ m}^3/\text{s}$

Fig 6a   Contours of radial velocity, impeller 1

Rotation

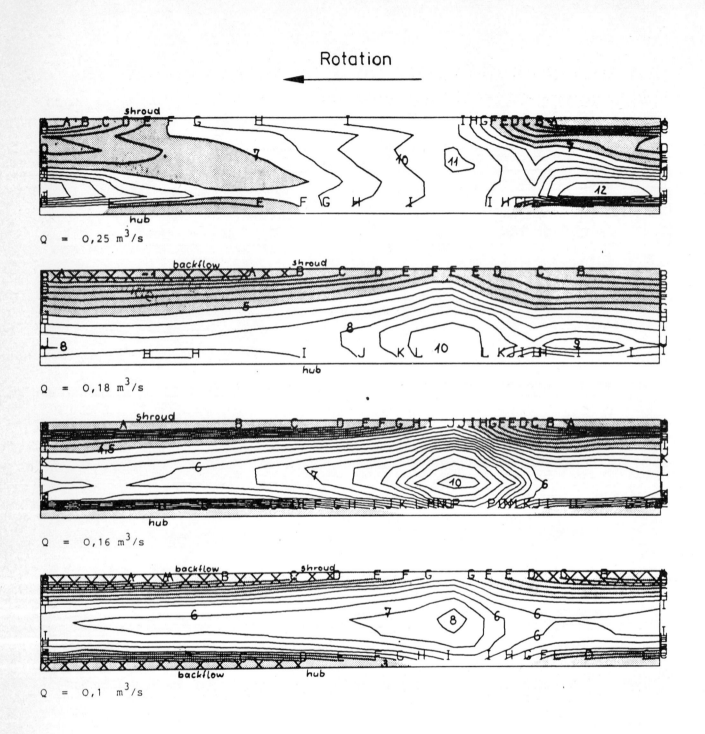

Q = 0,25 m³/s

Q = 0,18 m³/s

Q = 0,16 m³/s

Q = 0,1 m³/s

Fig 6b   Contours of radial velocity, impeller 2

C338/88 © IMechE 1988

C339/88

Flow analysis and part-load performance of model mixed-flow pumps

C CAREY, MA, PhD, S M FRASER, BSc, PhD, CEng, MIMechE, MRAeS and S SHAMSOLAHI, BSc, MSc, PhD
Department of Mechanical and Process Engineering, University of Strathclyde, Glasgow
D McEWEN, BSc
National Engineering Laboratory, East Kilbride, Glasgow

SYNOPSIS The flow in two model mixed-flow pump impellers operating in air has been examined using a laser Doppler anemometer, for a number of part load operating conditions. The experimental data have been examined critically using Euler's equation to explain differences in the overall performance characteristics of the two impellers.

NOTATION

g	acceleration due to gravity
H	impeller head
r	radius
U	impeller blade velocity
V_u	flow absolute tangential velocity
W_u	flow relative tangential velocity
β	blade angle
ϕ	flow coefficient
ϕ_0	best efficiency flow coefficient
ψ	head coefficient

Subscripts
1   impeller inlet
2   impeller outlet

1 INTRODUCTION

During the course of the last six years, an extensive and detailed programme of experiments has been carried out at the University of Strathclyde, on the flow behaviour in mixed-flow machines operating under design and off-design conditions. Laser Doppler anemometry has been used to obtain comprehensive sets of data describing the three-dimensional mean flow field in the rotating and stationary passages of the machines. The majority of these data have already been presented in the literature, and subsequently published in National Engineering Laboratory (NEL) Reports. This paper will give an overview of the experiments, and highlight the main findings with particular reference to the problems of 'stability' of part load performance characteristics.

2 EXPERIMENTAL FINDINGS

Two impellers were studied, operating in the casing shown in Fig 1. The Mark 2 impeller provides a continuously negatively-sloped head-flow curve, while the Mark 1 design produces a head-flow characteristic with a positive gradient in the range of 50% to 70% of design flow. The principal design features of the two impellers are given in Table 1, and the best efficiency point performance data in Table 2. The head-flow characteristics and the duty points at which the flow surveys were conducted are shown in Fig 2.

Table 1   Principal design features of impellers

		Mark 1 Impeller			Mark 2 Impeller		
		Hub	Mean-line	Tip	Hub	Mean-line	Tip
r_1	/mm	91	151	211	91	147	203
r_2	/mm	178	215	252	178	215	252
chord	/mm	244	275	290	316	333	357
β_1	/deg	37	31	29	26	30	30
β_2	/deg	46	33	27	33	27	25

Table 2 Best efficiency point performance of model pump with alternative impellers

	Mark 1	Mark 2
Flow coefficient	0.344	0.285
Head coefficient	0.280	0.243
Specific speed	2.33	2.37
Efficiency	0.87	0.83

In spite of the evident differences in performance, the experimentally determined flow fields within the impellers show a remarkable degree of similarity throughout the range of flows examined[1]. Indeed, at equal proportions of their respective best efficiency flow coefficients, the flow patterns in the two impellers appear almost indistinguishable.

Because of the limitations imposed on the length of this paper, it is not possible to include a full account of the results. However, for the benefit of readers unfamiliar with the work, a brief summary of the observations is given in the following paragraphs. Unless otherwise stated, the flow patterns described below apply to both Mark 1 and Mark 2 impellers. The changes in the flow patterns as the flow rate is reduced are summarised in Fig 3.

2.1 Best efficiency duty point, $\phi/\phi_0 = 1.0$

At best efficiency flow, the rotor blade passages are dominated by a large inviscid-type core flow. However, there is rapid growth of the casing annular boundary layer within the rotor, apparently precipitated by the blade tip leakage flow. The leakage jet rolls up into a vortex and separates from the blade suction surface, mixing with the mainstream flow, and spreading across the blade pitch. By the trailing edge of the rotor blades, the region of casing boundary layer and leakage vortex interaction extends across almost one quarter of the blade span.

2.2 Duty point A, $\phi/\phi_0 = 0.76$

As the flow is reduced from 100% to 76% of the best efficiency flow rate, the incidence of the flow to the rotor blades is increased by approximately 5 degrees across the full span of both rotors. The principal change in the observed flow pattern is an increase in the thickness of the casing boundary layer. Secondary flows are also greater in magnitude. However, in both rotors, the flow remains well-guided by the blades, with no large regions of separation, and no recirculation at inlet or outlet.

2.3 Duty point B, $\phi/\phi_0 = 0.59$

At this duty point, there is a marked change in the overall pattern of flow exhibited within the rotors. The main region of interest is at the rotor tip, where inlet recirculation and pre-rotation of the flow entering the rotor are evident. The pre-rotation is provoked by a region of separation on the pressure side of the blade, near the leading edge. This separation is noticeably stronger in the Mark 2 impeller than in the Mark 1 design.

Because the flow passage is partially blocked near the rotor inlet by this region of recirculation, the mean radius at which the flow enters the blade passage is reduced, and there is a spanwise movement of fluid towards the tip downstream of the blockage. This effect plays an important part in the head generation analysis carried out below.

2.4 Duty point C, $\phi/\phi_0 = 0.49$

In the tip region, the blade sections of both impellers are fully stalled. The extent of pre-rotation at the inlet tip increases, both in the magnitude of the tangential velocity upstream of the rotor, and in the thickness of the annular layer in which pre-rotation occurs (approximately 50% of the blade span). Recirculation in the tip region is also greater, giving a net reverse flow in the outer 10% of the blade span. This presents a greater blockage to the inlet flow than that described for duty point B, and hence there are increased amounts of spanwise flow downstream of the blockage. However, outlet recirculation is not established in either rotor.

2.5 Comment on trends exhibited by data

In view of the great similarity between the flow patterns observed in the two impellers, perhaps the most important outcome of direct examination of the flow patterns is that in this case, stability is not determined by particular flow phenomena, such as the onset of pre-whirl or outlet recirculation.

3 ANALYSIS OF THE DATA USING EULER'S EQUATION

In order to gain a better understanding of the differences in performance of the two impellers, a careful analysis of the experimental data has been carried out using Euler's equation. The analysis provides estimates of head generation by the separate mechanisms of "blade shape" work (ie. deflection or diffusion) and "centrifugal" work (caused by radial displacement of the flow in passing through the impeller). The mathematical basis for the analysis is as follows.

The well-known governing equation for the head generated by a pump is

$$gH = U_2 V_{u2} - U_1 V_{u1} \qquad (1)$$

Noting that the absolute tangential velocity V_u can be written as the difference between the blade velocity and the relative tangential velocity ie. $V_u = U - W_u$, this equation can be expanded to give

$$gH = (U_2^2 - U_1^2) - (U_2 W_{u2} - U_1 W_{u1}) \qquad (2)$$

The first term in the right-hand side of this equation results only from the radial shift of fluid streamlines in passing through the machine, since at any given radius, the blade velocity is constant. The second term is related to the change of relative tangential velocity which is caused by the blade shape. Effectively, the first term represents the head generated by centrifugal action and the second term is the head rise generated by diffusion.

The most common application of this equation is in the design of a pump. In the case of an axial machine, it is usually assumed that the centrifugal term is zero, and the magnitude of the diffusion term is estimated with the help of cascade data. For a centrifugal machine with purely radial blades, both terms must be considered, but the first term is easy to calculate because all of the fluid enters the impeller at one radius and leaves at a second radius.

For a mixed-flow machine, the picture is more complicated, because different streamlines enter and leave the rotor at differing radii. It is common practice nowadays, to use a potential flow solution obtained by computer-aided methods to estimate the effective radial displacement of individual streamlines passing through the rotor; the overall head rise across the impeller is then obtained by integrating all of the streamlines across the flow.

For off-design flows the picture is complicated still further by the fact that the fluid is distributed across the flow passages in a highly non-uniform manner. Regions of pre-rotation and recirculation cause large spanwise flows to occur, contributing to the difficulties. As a result, the displacement of individual streamlines within the impeller can not be readily calculated, and the off-design performance is unpredictable.

From the laser-Doppler measurements, the velocity distributions at inlet and outlet to the impeller have been determined to a high degree of accuracy. It was therefore possible to use the measurements to estimate the magnitude of each of the terms in equation (2) for individual streamlines passing through the impeller, and, by a process of integration, to determine the overall head generated by the separate mechanisms of centrifugal action and diffusion. This procedure was straightforward, but time-consuming, requiring much numerical integration and interpolation to estimate correctly the magnitude of radial displacement of streamlines at the outlet from particular radial positions at inlet.

Although it is not a new idea to separate the terms in Euler's equation in this way[2], reliable data on which to attempt an analysis of this type have not been available previously. Furthermore, the fact that the two impellers used in this experiment gave such similar patterns of flow behaviour makes the comparison between their respective Euler heads especially interesting.

4 RESULTS

The results of the Euler analyses are presented in Figs 4 and 5. Two pairs of curves are plotted in Fig 4, showing, for Mark 1 and Mark 2 configurations

(a) the measured head coefficient for the entire machine, including stator/diffuser;
(b) the total Euler head coefficient for the impeller only.

In Fig 5, curves (b) are repeated for comparison with two additional pairs of curves, showing

(c) the Euler head coefficient due to the 'centrifugal' action ie. the $(U_2^2 - U_1^2)$ term from Eqn (2);
(d) the Euler head coefficient due to the 'diffusive' action of the impeller blades ie. the $(U_2 W_{u2} - U_1 W_{u1})$ term from Eqn (2).

The data in graphs (b), (c), and (d) were estimated from the laser-Doppler measurements as indicated in section 3. The data in graph (a) were obtained from direct measurements of pressure rise across the entire machine at wall tappings.

4.1 Estimates of total Euler head

Consider first the graphs of total Euler head coefficient for the two impellers and the graphs of measured head coefficient (Fig 4). It is evident that the estimates of total head rise given by the Euler analysis are greater than the measured head coefficients. This result is expected, because energy is dissipated in the

stator due to aerodynamic losses. At the best efficiency point, the flow is well-guided by the stator blades, and the recovery of swirl energy is almost complete; hence the differences between the measured head coefficient and the Euler analysis estimates are small. However, as the flow coefficient is reduced, the incidence of the flow to the stator blades increases, eventually causing separation and stall. Under these circumstances, the stator losses are much larger; hence there is a greater discrepancy between the total Euler head rise across the impeller and the measured head coefficient across the entire machine.

The overall stability trends exhibited by the measured head coefficients are not fully reproduced in the total Euler head estimates. Whereas the measured head characteristic for the Mark 1 configuration shows a pronounced droop as the flow coefficient is reduced from 76% to 46% of the best efficiency flow, the Euler analysis of the same impeller flow shows a flat plateau in this range of flows. Thus, the dip in the head characteristic for the whole machine is the product of a flat impeller head curve combined with increasing stator losses as the flow coefficient is lowered.

For the Mark 2 impeller, the Euler head coefficient increases continuously as the flow coefficient is reduced. Furthermore, the rate of this increase always exceeds the rate of increase of stator losses, resulting in a 'stable' overall head-flow curve for the machine.

4.2 Mechanisms of head generation

In Fig 5, it is possible to examine the relative contributions of the different mechanisms of head generation in the two impellers. The uppermost pair of curves shows the characteristics produced by centrifugal action. It is immediately apparent that these curves are of almost identical shape; each point on the curve for the Mark 2 impeller is displaced vertically from the corresponding point on the Mark 1 curve by approximately the same amount (~0.07). The shape of the curves and the relative displacement of the Mark 2 curve can be explained qualitatively as follows.

At and near to the best efficiency point, it was found that the velocity distribution across the flow passage is reasonably uniform, and so for a particular impeller, the massflow-weighted mean inlet and outlet radii are almost independent of flow coefficient. Since the centrifugal action depends only on the radial position of streamlines at inlet and outlet from the rotor (Eqn 2), the head coefficient due to centrifugal action must also be constant. From Figure 5, it is clear that this observation appears to be valid for flow coefficients down to duty point A ($\phi/\phi_0 = 0.76$) at least. Once the onset of inlet recirculation occurs, the mean inlet radius is reduced, since in effect, the flow passage is partially blocked near the casing; however, the outlet is not blocked in the same way, and the mean outlet radius is almost unchanged from the higher flow coefficients. This explains why, as the flow coefficient is reduced to values below $\phi/\phi_0 = 0.6$, the curves begin to rise steeply. The spanwise displacement of the streamlines behind the inlet recirculation blockage increases with the size of the recirculation region. It was clear from the laser-Doppler measurements that the growth of the recirculation region was very similar in both impellers, and hence it is not surprising that the curves have nearly identical shapes.

The greater magnitude of the centrifugal head coefficient in the Mark 2 impeller over the Mark 1 impeller is the result of one of the main changes in the design. The tip of the Mark 2 blade was extended towards the bell-mouth inlet of the machine, reducing the mean inlet radius of the flow; the mean outlet radius was left unchanged. Therefore, the centrifugal term $(U_2^2 - U_1^2)$ is greater in the Mark 2 impeller, throughout the full range of duty points.

The two lowest curves in Fig 5 show the head characteristics due to the diffusive action of the two impellers. Note that for both impellers, this term is negative throughout the range of measurements, implying that $U_2 W_{u2} < U_1 W_{u1}$. Both curves show the same general trend; a head coefficient which rises steadily to a peak value as the flow coefficient is lowered from best efficiency duty, followed by a fall as the flow coefficient is lowered still further. This trend is due to the performance of the aerofoil blade sections as the flow coefficient is reduced (ie. as the incidence increases). Maximum 'lift' occurs at a particular incidence, above which the blade stalls. It should be remembered that for both rotors, the laser-Doppler results summarised in section 2 show clearly that stall does not occur simultaneously across the full span of the blades, but starts at the blade tip and spreads towards the root. This effect is due to the twisted blade geometry, which ensures higher incidence at the blade tip than at the root section.

It is evident that from Figure 5 that the diffusive contribution to the head coefficient in the Mark 1 impeller peaks at a higher flow coefficient than in the Mark 2 impeller. In addition, there is a larger drop in value of the Mark 1 coefficient as the flow is reduced from duty point A to C, than occurs at any stage in the Mark 2 coefficient. This large drop is the factor responsible for the droop in the overall head coefficient, because it takes place too rapidly to be compensated by the rise in centrifugal action described above. In the Mark 2 impeller, on the other hand, as the flow coefficient is lowered from duty point A to C, the rate of decrease of diffusive head is always less than the rate of increase of centrifugal head. In consequence, for the Mark 2 impeller, the total head curve for (representing the sum of the two contributions) displays a negative gradient throughout the range of flow coefficient.

5. CONCLUSIONS

Detailed velocity measurements have been obtained in the blade passages of two mixed-flow impellers operating under part-load conditions in the same model pump casing. The data have revealed remarkably similar flow patterns in the two impellers at corresponding fractions of their respective best efficiency duty points.

This result was in spite of the fact that the two impellers give quite different performance characteristics; the Mark 1 impeller produced a region of positive slope in the head-flow curve in the range 46% to 70% of best efficiency flow, while the curve for the Mark 2 impeller was of continuously negative gradient.

A better understanding of the nature of the performance differences between the two machines has been gained by using Euler's equation to analyse the experimental data. In this way, the relative contributions of the diffusive and centrifugal mechanisms of head generation in the machine have been separated and plotted. The centrifugal head characteristics of the two impellers are almost identical in shape. In contrast, the diffusion characteristics of the two impellers differ markedly; compared with the Mark 2 impeller, the Mark 1 impeller diffusion peaks at a higher fraction of its best efficiency flow coefficient, and then falls by a much greater extent when the blade stalls. This difference suggests that although the flow patterns in the two impellers appear to be very similar, there are significant differences in the aerodynamic performance of the blade sections. It is these subtle effects which must be predicted rather than the overall pattern of three-dimensional flow, if off-design performance is to be reliably determined using CFD methods.

ACKNOWLEDGEMENT

This work was sponsored by the Department of Industry and the paper is published with permission of the Director of NEL. The information contained in the paper is Crown Copyright.

REFERENCE

(1) CAREY, C., SHAMSOLAHI, S., FRASER, S.M. and WILSON, G. Comparison of the three-dimensional flow in two mixed-flow impellers. Proc. IAHR Symposium, Montreal, 1986, paper 48.

(2) WILSON, G. The development of a mixed-flow pump with a stable characteristic. NEL Report No. 110, 1963.

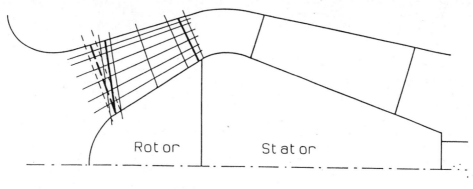

—— Leading edge of Mark 2 rotor blade
Traverse grid lines:
measurement points at intersections

Fig 1 Section through casing of model mixed-flow pump, showing LDA measurement grid

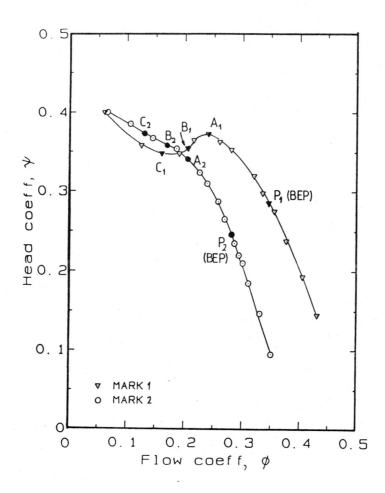

Fig 2 Performance characteristics of model pump with alternative impellers; black points indicate LDA flow survey duty points

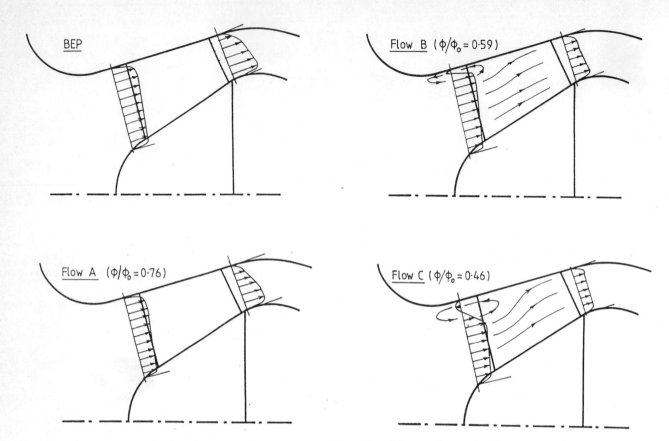

Fig 3   Summary of flow patterns observed at different duty points for
         both Mark 1 and Mark 2 impellers

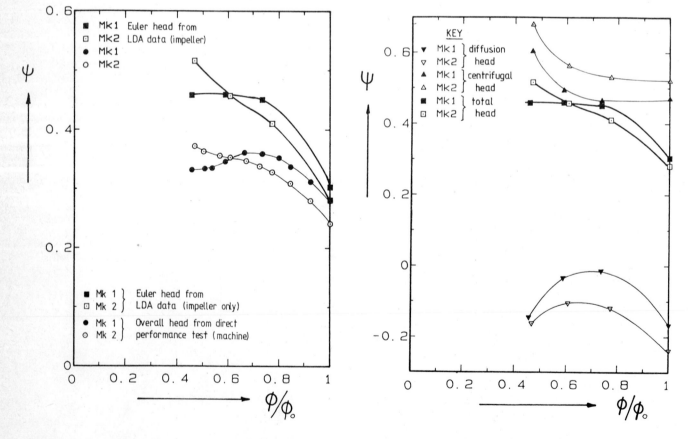

Fig 4   Comparison of head characteristics obtained from
         direct model testing with estimates using Euler
         analysis of LDA data

Fig 5   Head characteristics due to separate mechanisms
         of head generation

C340/88

Sewage pump efficiency and noise at part-load

T A McCONNELL
Weir Pumps Limited, Alloa

SYNOPSIS   The operation of rotodynamic pumps at part load can result in generation of hydraulically initiated noise caused by turbulent flow regimes within the waterway passages. This factor can be particularly prevalent in sewage pumps where the impeller and diffuser geometries are generally compromised by the requirement to pass solids of specified dimensions.   This paper reviews design features which affect noise generation and shows that good part load noise performance can be achieved with minimal compromise to pump efficiency.

NOTATION

$$K \; = \; \frac{2\pi \; NQ^{0.5}}{(gH)^{0.75}} \qquad \text{Type Number}$$

N = speed of rotation Revolutions per Second

Q = volume rate of flow Metres per Second

g = gravitational acceleration Metres per Second^2

H = head Metres

1   INTRODUCTION

The operation of rotodynamic pumps at part load can, as is well known, result in the generation of hydraulically initiated noise caused by turbulent flow regimes within the waterway passages.   This feature can be particularly prevalent in sewage pumps where the impeller and casing volute geometries are generally compromised by the requirement to pass solids of specified dimensions.

Elimination of most noise can readily be achieved by using fully concentric volutes.   This solution does, however, have a significant detrimental effect on pump efficiency at normal duty which in todays energy conscious market is not acceptable. The use of a spiral volute casing with close cutwater clearance will maximise efficiency but may result in a pump which is noisy at part load.

The design engineer requires to produce a pump design which minimises noise generation with least compromise to pump efficiency.   This paper reviews design features which affect noise generation and shows that good part load noise performance can be achieved with minimal efficiency reduction.

2   SOURCES OF HYDRAULIC NOISE

Generation of hydraulic noise in rotodynamic pumps is generally related to:-

(a) Impeller vane and diffuser cutwater interaction effects.
(b) Impeller inlet cavitation.
(c) Impeller recirculation at part flows.

Generally, though not always, the source of the noise in sewage pumps can be ascertained by carefully listening to the pump.

Interaction noise is loudest close to the casing cutwater.   Also on some units this noise can give the impression of intermittently moving around the volute. The noise normally occurs at low flows and disappears as flow is increased.

The noise generated by interaction of impeller and volute cutwater is sometimes confused with inlet cavitation which is either due to lack of nett positive suction head or inlet recirculation.

With inlet cavitation, noise is loudest at the impeller eye and the noise generally increases with flow until severe loss of head generation occurs due to lack of nett positive suction head.

On occasions when extensive inlet recirculation noise occurs determination of the noise source can be more difficult to ascertain due to the pervasive nature of the noise.   Recirculation and interaction noise both reduce with increasing flow and increasing nett positive suction head.

Use of noise measurement instruments will establish the source.

The main source of noise in sewage pumps is usually associated with impeller vane and volute cutwater interaction effects which are a result of compromised hydraulic design to satisfy solids handling requirements.

Sewage pumps are required to pass solids the size of which are invariably specified by the customer as being spheres of a certain diameter, usually 75mm or 100mm equivalent diameter. The pump manufacturer requires to ensure, and may be required to demonstrate, that the specified sphere will in fact pass through the pump. On figure 1 a cross section of a typical sewage pump is illustrated.

In achieving this solids handling requirement the designer is forced into producing a design which contains inherent compromises with respect to optimum flow conditions within the pump (1). It is necessary to incorporate features, the most significant of which is fewer than preferred impeller blade number, which result in reduction in efficiency due to significant additional shock, flow slippage and recirculation losses.

Increased turbulence in the flow regimes within the impeller and volute can result in intense vortices being formed. When attempting to maximise efficiency by minimising cutwater clearance these vortices can become vigorous enough at part load to form vapour cores at their centre. When pressure at the vortex centres recovers above vapour pressure the vapour cores collapse violently exhibiting typical cavitation noise.

Noise generation in sewage pumps resulting from impeller blade and cutwater interaction, except in very severe cases, is not damaging to the pump. However the noise can, particularly if pumps are operated for significant periods at part load where noise is most significant, be environmentally unacceptable. Designing for quiet operation, or at least minimal noise, is therefore never the less an important consideration.

3 NOISE VERSUS FLOW RELATIONSHIP

The majority of sewage stations are designed such that the pumps always operate with a positive suction head.

The pump is normally started and stopped using level control in the suction sump. On the discharge side the pump discharges via a non-return valve into the system. With the exception of maintainance operations the suction and delivery isolation valves are always left fully open.

On Figure 2 the operational sequence is shown for single pump operation in a typical system. The arrows trace the locus of pump operating point from start up, through normal pumping, to shut down.

On start up the pump operates at zero flow until sufficient head is obtained to overcome system static head, flow then being delivered to the system as the non return valve opens.

As the suction sump is emptied the static head increases up to the maximum value and the flowrate reduces to the minimum. Thereafter on shut down the pump flowrate reduces until the non-return valve shuts and the pump eventually comes to rest.

Superimposed on this figure is the pattern of noise generation which can be experienced on a sewage pump designed for maximum efficiency using a tight cutwater clearance. From this diagrammatic example it can be seen that several noise regimes can exist, viz:-

(a) Noisy - an area of operation can exist near zero flow rate at speeds close to normal operational speed where noise can be heard as continuous crackling and banging.
(b) Moderate Noise - as the flowrate increases or speed decreases the noise level within the unit will decrease in intensity. The noise may still be continuous at lower flowrates becoming intermittent as flowrate increases.
(c) Intermittent Minor Noise - in this region the noise is sporadic and will generally be heard as random minor crackles.

In the example illustrated although the pump may exhibit undesirable noise characteristics at very low flowrates the unit may be perfectly suitable for operation with the typical system indicated. Minor noise emission would be apparent as the static increases and the flowrate reduces but in the normal flow range the pump would be quiet.

The example shown is not representative of all installations. The range of static heads will often be greater with consequently lower minimum flowrates where the operational zone would extend into the noisier regimes.

In todays markets, whilst initial cost remains crucial, operational costs and in particular energy costs, take on over increasing importance. Maximising efficiency has however to be tempered in sewage pumping applications by proper awareness of the undesirable environmental effects of noise.

The correct approach for pump design is to ensure quiet operation in the normal flow range and to accept minimal noise at low flow rates such that pump efficiency is least compromised.

4 DESIGNING FOR OPTIMUM PERFORMANCE

When designing a sewage pump the designer in addition to achieving the customers head and flow has to fulfill the essential requirements of ensuring that the pump will allow the specified sphere size to pass through the impeller and casing.

In all conventional volute diffuser type pumps the restriction on sphere size is to be found within the impeller waterway passages. Figure 3 illustrates a typical sewage pump impeller. Generally two areas cause the restriction on sphere size:-

The Impeller Inlet - the inlet area into the impeller waterways often contains the smallest dimension between flow passage surfaces. A rough guide to the sphere size is obtained by the design engineer looking at area normal to flow at the inlet. However, the actual ability to pass a sphere is a three dimensional problem and traditionally has best been finally checked out on a model or the pattern. These checks can now be made using three dimensional computer aided draughting systems.

The Impeller Outlet - as pump generated head increases the outlet width to pass a given flowrate decreases and therefore on higher head (lower type number) sewage pumps the problem of sphere handling can be most significant at the outlet of the impeller.

On large sewage pumps sphere size is not generally a problem since the impeller waterway passage areas are large enough to pass specified solids. The pump impeller and volute hydraulic design in these cases tends towards conventional fresh water pump practice.

To enable the pump designer to meet the criterion of sphere size on a specific design it becomes necessary, except on large pumps, to move away from optimum design practice. As stated previously, the design will be less efficient than a normal pump designed for the same output.

The deviations from optimum practice, all of which contribute to generation of part load noise, are as follows:-

(a) Number of Impeller Vanes
The number of impeller vanes must be reduced thus resulting in high blade loading which can often be 2 to 3 times, that of normal design practice.

The use of fewer blades leads to increased recirculation, slip at vane outlets and increased pressure pulses. Where very low vane numbers are used, i.e. 2 or 3, efficiency and suction performance can be significantly affected.

(b) Impeller Inlet Diameter
An increased diameter impeller inlet often has to be used to pass the required sphere size. This leads to increased recirculation within the impeller particularly at the inlet, and impellers can often be recirculating at the duty flow. Efficiency levels can be significantly reduced due to large impeller eye diameters.

(c) Impeller Vane Design
Requirement for increased mechanical strength to cater for shock loads caused by solids being pumped, and higher loadings due to fewer vanes, necessitates that vanes be thickened. This results in increased shock losses at inlet and outlet, with increased pressure pulses at outlet due to vane wake effects. Effects on pump efficiency vary with unit size and are significant on small units but negligible on large units.

Vane design often requires to be adjusted to give a less than ideal vane angle progression at the inlet to make distance between vanes greater. This results in marginal efficiency reduction and increased inlet recirculation.

(d) Casing Volute
The design of the volute is subject to the limitations required to pass the specified sphere size coupled with the need to have a robust cutwater which will not be damaged by solids passing through the pump. The selection of cutwater clearance significantly influences part load noise and peak efficiency.

It is not considered necessary on sewage pumps to design the volute to allow the specified sphere to pass between the impeller and the cutwater since the solids are considered to be of fabric in nature and relatively soft. On certain slurry or sugar beet pumping applications it is considered essential to allow for hard spheres of specified diameter to pass between the cutwater and impeller thus minimum cutwater clearance then needs to be greater than sphere size.

The generation of noise in sewage pumps is related in combination to all of the above features and mostly results from the interaction between the impeller vanes and casing cutwater. On Figure 4 a diagrammatic view is shown of typical flow regimes. As flow leaving the impeller is distorted due to the cutwater the vortices are "squeezed" and the core accelerates. The pressure in the centre of the core drops to vapour pressure and vapour pockets are formed. When these collapse in the diffuser cavitation type noise is produced.

Experience has shown that in almost all instances the collapse of bubbles occurs in the water stream rather than on the metal surfaces. Thus whilst the noise may be objectionable no damage is done to the casing material.

This has been verified by the authors company using paint erosion techniques developed for studying cavitation erosion on impeller inlet designs. Several casing volutes were painted with specially developed paint and the pumps then run under noisy part load operating conditions.

After test the units were stripped and checked for evidence of paint removal.

In designing for optimum performance it is important the relationships between features of design are understood. Design methods need to be established so that optimum performance can be confidently predicted for all of a range of pumps to cover the range of flows and generated heads required by the market. The authors company has used subjective noise information from production performance tests on its traditional range of sewage pumps and also data from prototype tests to establish design criteria which have since been applied to new designs and retrospectively to modify existing designs with good result. Some of the performance relationships identified from the tests are presented in the following section.

5 PERFORMANCE RELATIONSHIP ESTABLISHED FROM TEST DATA

5.1 Production Tests

To provide data for analysis it was decided for simplicity to monitor all production sewage pumps over a period of time on the workshop test bed.

After some initial trials with standard noise meters surveying the noise characteristics of the standard range pumps it was decided that the most cost effective means was subjective assessment by the design teams.

This decision was made since the response speed of standard noise meters (2) was unacceptable for quantifying peak noise values particularly when measuring intermittent noise. The readings obtained did not match well with the impression of noise intensity observed by the engineers.

The use of more sophisticated equipment was discounted since this would lead to unacceptable delay and additional costs on the test bed. Since our concern was for clearly audible noise the subjective evaluation approach was considered adequate and acceptable.

Initially a system with five noise categories was used. This, however, proved to be too many for subjective evaluations so the number of categories was reduced to three.

Noisy:- Noise levels objectionable to listener at closed valve. Moderate noise at part flow.

Moderately
noisy:- Some noise at closed valve but not objectionable. Little noise at part flow.

Quiet:- No noise of any note.

To maximise and speed data collection, where production test commitments permitted maximum diameter impellers were fitted to production units for initial test.

From the data collected various plots were prepared to establish feature relationships as they appeared to influence noise.

Cutwater clearance was known to have the most significant effect on sewage pumps and it was known that pumps could always be made quiet if the clearance was increased. Data in reference (3) showing variations in cutwater clearance for a normal design 8 bladed impeller is not representative of the variations that can exist on impellers with lower vane numbers. Since cutwater clearance was known to have the most significant effect, various parameters were reviewed against this feature.

It had been intuitively thought that vane number in conjunction with cutwater clearance would indicate a direct relationship with noise. On Figure 5 results are indicated, and show, contrary to expectation, that no clear trend exists.

Consideration of other factors such as generated head, blade lift coefficient, vane inlet design including recirculation also showed no identifiable trends. As mentioned previously, impeller overall blade loadings can be 2 to 3 times higher than in normal pump designs and it was surprising not to see a relationship between these factors.

Since it was known that reduction in diameter or running speed reduced and eventually eliminated noise it was decided to examine the influence of differential pressures.

Since the power absorbed at say 100% and 95% diameter could be determined then the power absorbed by the outer 5% of the impeller could be ascertained.

Knowing the impeller outlet dimensions and vane number and assuming for simplicity that the difference in power was directly related to torque on the pressure face of the impeller vanes, a vane tip differential pressure could be calculated.

A plot of this is shown on Figure 6 and it shows that there is a relationship in that high tip differential pressures require larger cutwater clearances.Anomalies which existed in this relationship were shown by further analysis to be related to the casing sidewall diffusion angle or to large eye diameter impeller design.

Casing sidewall angle contributes significantly to noise since as this angle increases so does the tendency for greater flow separation from the diffuser side walls refer Figure 7. Consequently this leads to greater turbulence and potential for part flow noise.

Low limiting eye velocities, resulting in relatively large eye impeller designs can have a detrimental effect on both noise and efficiency performance.

With regard to ragging and choking it has long been believed that some recirculation in the eye of an impeller adjacent to the shroud alleviates potential problems. Reverse flows tends to remove the rags from impeller vane inlets carrying them back into the suction. Thereafter the rags may re-enter the eye passing through the pump.

On larger sewage pumps blockages are not a significant problem (4) and even on medium sized units this aspect can however be overcompensated for by specifying restrictive velocities. Due to the design requirement for solids handling, sewage pump impellers invariably recirculate at inlet close to best efficiency flow and tend to self clean away.

Designing for unnecessarily low eye velocities can lead to loss of 2 or 3 efficiency points and impaired part load noise characteristics. Increase in eye diameter causes loss of vane area and promotes the tendency for excessive impeller inlet and exit recirculation.

From this analysis a relationship between volute side wall angle, vane tip pressure differential, and volute cutwater clearance was established for noise and efficiency performance. This relationship is summarised in Figure 8 for impellers which have straight radial vanes at outlet.

Subsequent testing has shown that this relationship may be slightly modified and pump performance further improved by special contouring of the volute cutwater and skewing of the impeller vane tips.

5.2 Laboratory Tests

On volute type pumps the efficiency drops off as the impeller diameter is reduced mainly due to casing recirculation losses. In an ideal volute no flow would pass between the impeller and cutwater more than once. As the diameter is reduced however flow is carried round the volute greater distances before being discharged.

In a similar manner if an impeller is tested with different casings of varying design where the cutwater clearance is increased the efficiency falls off.

Laboratory tests were carried out using three different casing designs for a pump with Type Number K = 2.2.

(a) Spiral volute.
(b) A 65° partially concentric diffuser.
(c) A 220° partially concentric diffuser.

On Figure 9 a plot is shown of drop in efficiency versus degree of casing concentricity where the degree of

concentricity is defined as being the angle measured from the cutwater round the diffuser over which the diffuser wall is concentric to the impeller. Thereafter the diffuser expands as a normal spiral volute.

From this curve it can be readily observed that cutwater clearance and degree of casing volute concentricity has a marked effect on pump efficiency. It is important to be able to accurately determine the required cutwater clearance for any unit such that it will give satisfactory noise performance at part load flows without unnecessarily compromising pump efficiency.

6 CONCLUSIONS

The satisfactory operation of a sewage pump with respect to minimising noise whilst maximising efficiency is primarily influenced by cutwater clearance, impeller blade tip differential pressure and diffuser sidewall angle.

With established data on feature relationships and careful attention to feature design, optimum performance with respect to noise generation and efficiency can confidently be predicted.

Low limiting eye velocities sometimes specified for sewage pump designs, on the premise that this alleviates ragging problems can have a detrimental effect on both noise and efficiency performance and the benefit of these low velocities should be questioned and studied with respect to actual site operating experience.

REFERENCES

(1) Anon Pumps, Pompes, Pumpen 1969, P447-8, 450, 452

(2) Lightfoot, P.W.,Airborne Noise of Pumping Plant, The South African Mechanical Engineer, 1977, Vol. 27, P144-150

(3) Clark, T.A., Hydraulic Noise Generation Its Detection and Uses as a Diagnostic Tool, BHRA Report RR1104, 1971

(4) Johnson, M., Pumping Sewage and Sewage Sludge, Effluent and Water Treatment Journal, 1980, P575-578

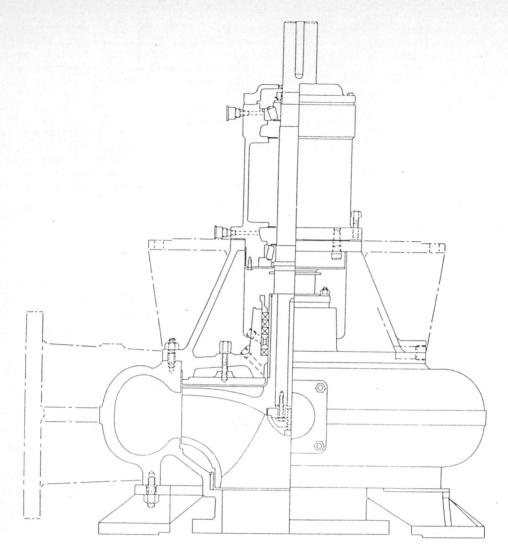

Fig 1   Cross-section of typical sewage pump

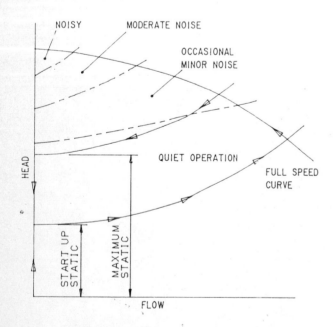

Fig 2   Comparison of operational condition and pump
noise characteristic for an efficiency-optimized
sewage pump

Fig 3   Typical sewage pump impeller

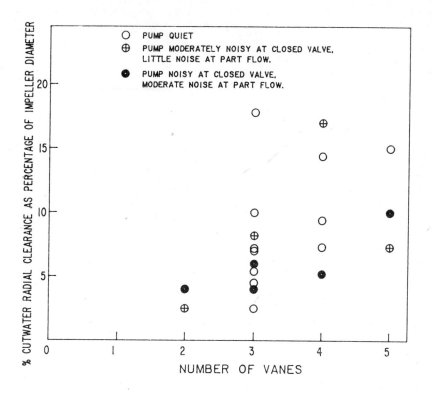

Fig 5   Noise ratings at various cutwater radial clearances and blade numbers

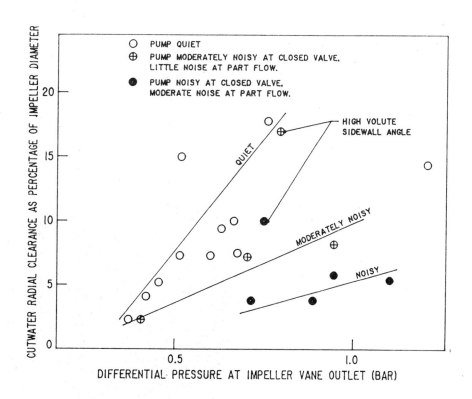

Fig 6   Noise ratings at various cutwater radial clearances versus closed-valve
impeller vane differential pressures

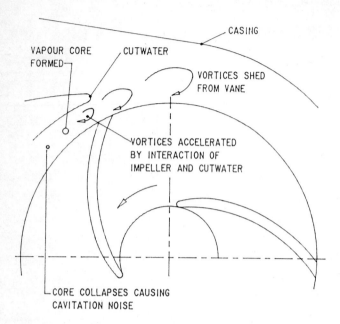

Fig 4   Flow regime at impeller outlet when interaction
         occurs with cutwater

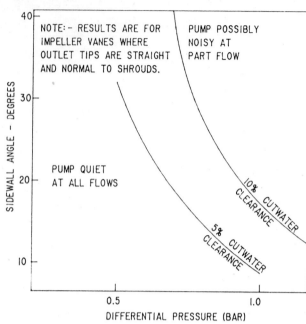

Fig 8   Limitations of vane-tip pressures for quiet operation
         at varying casing sidewall diffusion angles

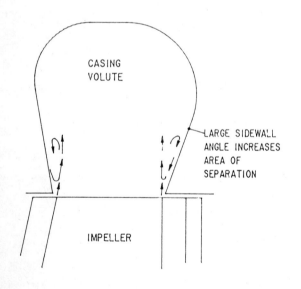

Fig 7   Flow pattern; large volute sidewall angle

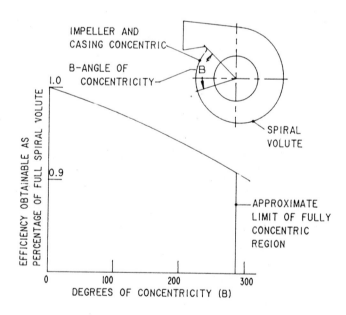

Fig 9   Comparison of pump efficiency for full spiral and
         partially concentric casings; for pump of type
         number K = 2.2

C341/88

The influence of occluded or dissolved air contained in water on pump cavitation

B LE FUR, J F DAVID and D PECOT
Centre Technique des Industries Mecaniques, Nantes, France

ABSTRACT

This paper sets out test results showing :

- first, the influence of occluded air in water (volume percent ranging from 0 to 5 %)

- second, the influence of the oxygen content dissolved in water on the cavitation behaviour of 2 pumps with a specific speed of 20 and 80.

ABBREVIATIONS

D_N : nominal diameter

N   : rotational speed (r. p. m.)

Q   : flowrate

H or Hmt : total manometric head

NPSH : net positive suction head

NS : specific speed (or SS)

$$(NS = \frac{N(r.\,p.\,m.)\,Q^{1/2}(m^3/s)}{H(m)^{3/4}})$$

1. REASON FOR THE INVESTIGATION

In 1983 - 1984 following a request from pump manufacturers, the CETIM of Nantes conducted a bibliographic research focused on the theme :

" *Influence of the following parameters :*

- *geometric scale*
- *rotational speed*
- *air content*
- *thermodynamic effect*

on pump cavitation behaviour ".

This research was aimed at setting the outlines of experimental investigations.

After the bibliography produced was synthetized and analyzed, it was decided in accordance with the pump industry, to carry out experimental researches about the influence of occluded and dissolved air content on pump cavitation.

2. PUMPS PUT TO TRIAL

The experimental program has been conducted using two pumps with a specific speed of 20 and 80 respectively.

The NS 20 pump impeller diameter is 330 mm, inlet diameter is 133 mm and the impeller features 10 blades. Nominal diameters at discharge and suction ends are 125 mm and 100 mm respectively.

The NS 80 pump impeller diameter is 345 mm, inlet diameter is 74.5 mm and the impeller features 6 blades. Nominal suction and discharge end diameters are 300.

3. INFLUENCE OF OCCLUDED AIR

3.1 Test loops

To study the influence of occluded air on pump cavitation the pumps are mounted on to an opened system loop.

In both cases, air is injected through a grid showing a great number of 0.2 mm D pores located at 1 suction D_N from the pump, upstream the suction flange. The sleeves on either side of the grids are made of plexiglass.

NS 80 Pump

- The suction line is fitted with a control valve very much downstream the pump and with 300, 400, 500 and 600 D_N pipings.

- The discharge line features 300 and 400 D_N pipings and a 2000 m3/h screw type flow meter.

- The pump is operated by a 460 KW d. c. motor.

NS 20 Pump

- The suction line is fitted with a control valve located very much downstream the pump and with 125, 150 D_N pipings

- The discharge line is fitted with 100 and 150 D_N pipings and with a 450 m3/h electromagnetic flow meter.

- The pump is driven by a 50 KW d. c. motor.

3.2 The trials

Due to water quality considerations, water level was lowered for the performance of the cavitation trials. Preliminary tests through valve regulation at the suction end demonstrated that valve cavitation was felt at pump suction end and that a non-negligible and incomputible amount of gaz was present. This is why we decided to lower water level in the test tank

NS 80 Pump

- The tests are carried out at three different rotational speeds :

$$
\begin{aligned}
&723 \text{ r.p.m.} \quad (Q = 820 \text{ m3/h}) \\
&912 \text{ r.p.m.} \quad (Q = 975 \text{ m3/h}) \\
&1150 \text{ r.p.m.} \ (Q = 1240 \text{ m3/h})
\end{aligned}
$$

- With flowrate measured at 7 points for every speed :

$Q/Q_N = 0.2 - 0.4 - 0.6 - 0.8 - 1 - 1.1 - 1.25$

(The nominal flowrate point is the pump maximum efficiency).

- And with 5 values of air volume percent referred to pump suction pressure for each point :

$\alpha = 0 \% - 1\% - 2 \% - 3.5 \% - 5 \%$

NS 20 Pump

- The tests are carried out at 1500 r.p.m. ($Q_N = 142.5$ m3/h) with flowrate measured at the following points :

$Q/Q_N = 0.6 - 0.8 - 1 - 1.10$

- Each point is put to test for 5 values of the air volume percent (referred to pump suction pressure) :

$\alpha = 0 \% - 1 \% - 2 \% - 3.5 \% - 5 \%$

3.3 Measurements

Most measurements are made using automatic data collection otherwise they are keyed manually into the computer.

They are immediately processed by the computer which yields :

- Pump conventional characteristics such as rotational speed, torque, flowrate, head, power, efficiency, upstream pressure and the available NPSH),

- The values necessary to determine the volume percent of injected air such as air temperature, air injection pressure, ΔH of the Pitot tube.

- The volume percent of injected air referred to pump suction pressure, and adjustement data if required.

3.4 Measurement and computation of the volume percent of injected air

The volume percent of injected air referred to pump suction pressure is equal to the ratio of the flowrate of injected air referred to the pump suction pressure to water flowrate, multipied by 100 %.

The injected air flowrate is measured by a Pitot tube fitted to the air injection line to the grid.

The tilted sight glass reads the injected air flowrate at air injection pressure and injected air temperature.

The injected air flowrate at pump suction pressure and at water temperature can be easily obtained from the known water temperature and pump suction pressure. To obtain the percentages : 1 % - 2 % - 3.5 % and 5 %, air flowrate is adjusted using injection pressure. The Pitot tube should be calibrated before this adjustment is made.

3.5 Results

For each rotational speed the results are given as the change in the 3 % NPSH (i. e. NPSH value for a 3 % decrease in the total manometric head) in terms of the Q/Q_N ratio for all 5 air percentages.

The datum height used to compute the NPSH for a given air percentage is the total manometric head obtained from this very air percentage.

The change in the 3 % NPSH curve = f (Q/Q_N) in terms of the volume percent of injected air ($\alpha = 0 - 1 - 2 - 3.5 - 5$ %) is illustrated in Figure1 (723 r.p.m.), Figure 2 (912 r.p.m.) and Figure 3 (1150 r.p.m.) for the NS 80 pump and in Figure 4 (N = 1500 r.p.m.) for the NS 20 pump.

For large flowrate generally associated with high air contents, a few points are missing because the NPSH were too high. (The amount of NPSHs available is restricted because the trials are conducted using an open system loop).

Small flowrate points (0.2 QN for the 3 speeds and 0.4 QN at 723 r.p.m. for the NS 80 pump, and 0.2 QN and 0.4 QN for the NS 20 pump) could not be investigated due to the excessive recirculations which made it impossible to adjust the amount of injected air.

The change in total manometric head for the highest level in the tank in terms of injected air percentage is shown in Figures 5, 6 and 7 for the NS 80 pump and in Figure 8 for the NS 20 pump.

These results show that the behaviour of the NS 80 pump (and especially the total manometric head), excluding cavitation, is first affected in high flowrates for air percentages about 2 - 2.5 % approximately. The group of results about 0.8 QN is then affected but for higher air percentages (3.5 - 4 % approximately). About 5 % the total manometric head of all the flowrate points is on a downward trend.

On the contrary, the cavitation behaviour of the NS 20 pump seems to be affected at lower air percentages. A percentage of air ranging from 1 to 1.5 % may modify the NPSH when Q/QN is about 0.8 or 1. The NPSH is not modified at lower flowrates, i.e. when Q/QN is about 0.4 - 0.6, until air percentages of 2 - 2.5 % approximately are reached.

It can thus be concluded that the NPSH of the NS 20 pump seems to be less dependent on the occluded air content than the NS 80 NPSH (curves are closer). Also the NS 20 pump total manometric head change is smaller than the NS 80's.

It should be emphasized that for both pumps, the point at 0.6 QN is not so affected by the injected air content as the remaining flowrate points which may be due to the recirculations occuring in the impeller for this flowrate.

4. INFLUENCE OF THE DISSOLVED AIR CONTENT IN WATER

4.1 Test loops

The pumps are mounted on to a closed system loop to investigate the influence of the dissolved air contained in water on pump cavitation.

NS 80 Pump

- The closed system loop includes D_N 300 and 400 pipings, a steady flow condition vessel and a pressure variator.

NS 20 Pump

- The closed system loop consists of D_N 100 - 125 and 150 pipings, a steady flow condition vessel and a pressure variator. It also includes a valve and a 450 m3/h electromagnetic flowmeter and a parallel mounted water de-aerating unit. Water taken from the loop by an hydroejector is then sprayed above water level in the pressure variator.

The sleeves at the pumps suction ends are made of plexiglass.

The dissolved oxygen content used to determine the amount of air dissolved in water is permanently read by an oxymeter.

The oxymeter is fitted to a branch pipe between the suction and the discharge line. The measurement accuracy is periodically checked against chemical measurements.

4.2 Trial performance

Cavitation tests are carried out by depressurizing the loops.

NS 80 Pump

- The trials are conducted at 3 rotational speeds :

> 912 r.p.m. (QN = 975 m3/h)
> 1150 r.p.m. (QN = 1240 m3/h)
> 1438 r.p.m. (QN = 1550 m3/h)

- For each speed, flowrate is measured at 7 points corresponding to the ratios :

Q/QN = 0.2 - 0.4 - 0.6 - 0.8 - 1 - 1.13 - 1.25

- For each point, the tests are carried out for 3 values of the dissolved oxygen content, as a minimum.

NS 20 Pump

- The tests are carried out at three rotational speeds

> 875 r.p.m. (QN = 83.1 m3/h)
> 975 r.p.m. (QN = 92.6 m3/h)
> 1470 r.p.m. (QN = 139.6 m3/h)

- For each speed, flowrate is measured at 5 points corresponding to the ratios :

Q/QN = 0.615 - 0.835 - 1 - 1.32 - 1.55

- For each point, the tests are carried out for 3 values of the dissolved oxygen content.

The content of dissolved oxygen decreases as pressure falls. For every cavitation test, the quantities recorded are : initial content of dissolved oxygen, oxygen content for the 3 % NPSH and oxygen content at the termination of the test.

The oxygen content for the 3 % NPSH is used to calculate the change in the NPSH.

4.3 Results

NS 80 Pump

- The results are given per rotational speed as the 3 % NPSH change in terms of dissolved oxygen content read in this point, with a curve for each flowrate curve.

Figure 9 : N = 912 r.p.m.
Figure 10 : N = 1150 r.p.m.
Figure 11 : N = 1438 r.p.m.

At 912 r.p.m. and 1150 r.p.m. the point on the curve with a circle round is the point obtained using the opened system loop (assuming that water was then saturated with dissolved air).

For NPSH comparatively small values (less than 4 m approximately) it was technically impossible to obtain dissolved oxygen contents higher than 5 mg/l, due to the following reason : as the test loop is relatively large-sized, pressure reduction using pumps was quite a slow process and the water had enough time to de-earate. This is the reason why in these circumstances the dissolved oxygen content is fairly small.

Curves in Figures 9, 10 and 11 show that the NPSH time history depends very much on the dissolved oxygen content ; however, it becomes more steady when oxygen content is under 5 to 6 mg/l/ The change seems to be all the more sensible since the NPSH is small.

The change in NPSH with the content of dissolved oxygen accounts for some differences found between NPSH closed system loop testing in the laboratory and pump operation on the site, after it was mounted on to an opened circuit (NS 80 pump).

NS 20 Pump

- test results are shown in Table 1.

No patent industrial NPSH (i. e. 3 % NPSH) change could be related to the dissolved oxygen content for this centrifugal pump (NS 20).

NPSH curves in terms of flowrate obtained from the 1500 r.p.m. tests conducted both with an opened and an a closed system loop are similar.

4.4 Conclusion

The tests allow to quantify the influence of existing occluded air on the NPSH of different types of pumps and to find that the influence of the dissolved oxygen content depends on the type of the pump used : it may be considered as negligible for centrifugal pumps (NS 20) while it may become very sensible for mixed-flow pumps (NS 80). In the latter case, always remember that a pump may feature dissimilar NPSHs depending whether it is connected to an openened or to a closed system loop.

REFERENCES

F.G. HAMITT
Effects of gas content upon cavitation inception, performance and damage
Journal of Hydraulic Research - Vol 10 - n° 3 - 1972
p. 259- 290

I.J. KARASSIK
Centrifugal pump clinic 1/83
World pumps - Nr 2 - Février 1983

MURAKAMI
MINEMURA
Effects of entrained air on the performance of Centrifugal pumps under cavitating conditions
Bulletin of the J. S. M. E.
Vol 23 - n° 183 - 1980

TSAI
Accounting for dissolved gases in pump design
Chemical Engineering July 26, 1982
Vol. 89 - n° 15

J. GÜLICH
Grandeurs caractéristiques de la similitude pour la capacité d'aspiration et l'extension des bulles dans les pompes
Revue Technique Sulzer - 2/1980

Table 1   Influence of the dissolved oxygen content on the 3 per cent NPSH of a pump with a specific speed of 20

Speed : 875 tr/mn

	Q/QN 1,55			Q/QN 1,325			Q/QN 1			Q/QN 0,835		
Initial O^2 content	10,3	6,7	4,6	10,4	7,2	3,5	10,3	8,6	4,6	10,3	6,9	3,8
O^2 content at 3 % NPSH	9,7	6,7	4,6	7,8	6,4	3,5	≈5	≈4	≈3,5	4	3	2,2
Final O^2 content	8,7	6,6	4,6	6,5	6,2	3,5	4,2	3,6	2,6	2,8	2,3	1,8
NPSH 3 % (m)	7,50	7,53	7,53	4,12	4,05	4,12	1,53	1,51	1,43	1,16	1,18	1,17

Speed : 975 tr/mn

	Q/QN 1,55			Q/QN 1,325			Q/QN 1,04			Q/QN 0,835			Q/QN 0,615		
Initial O^2 content	10	5	3,2	10	7,2	4,6	10,3	7	4,1	10,2	6,2	3	10,2	5,5	3,2
O^2 content at 3 % NPSH	9,6	4,9		7,7	7,1		5,7	4,4	3,2	2,8	2,3	1,6	1,2	1,2	1,3
Final O^2 content	8,6	4,7	3,1	5,2	4,9	4,3	4,4	3,2	2,05	2,4	1,8	1,4	0,8		
NPSH 3 % (m)	9,95	9,95	9,88	5,81	5,82	5,80	1,99	2,09	2,05	1,13	1,14	1,17	0,87	0,90	0,90

Speed : 1475 tr/mn

				Q/QN 1,32			Q/QN 1,04			Q/QN 0,835			Q/QN 0,615		
Initial O^2 content				8	6,8		8,6	8,1	5,7	9,2	8,7	4,7	10,9	7,8	4,7
O^2 content at 3 % NPSH				8	6,8		6,3	5,6	5,2	2,4	3,5	2,5	4,2	3,6	2,6
Final O^2 content				8	6,7		4,7	4,2	4,4	1,4		1,3	3,3	2,4	2
NPSH 3 % (m)				12,85	12,70		4,66	4,45	4,46	2,22	2,18	2,23	1,96	2,05	1,90

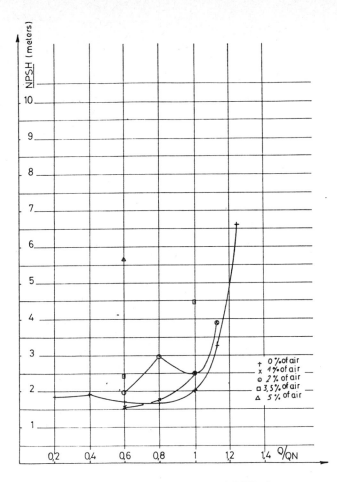

Fig 1   Influence of occluded air on the NPSH of a pump with a specific speed of 80 (N = 723 r/min)

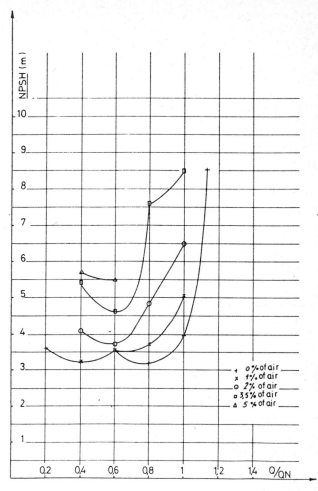

Fig 3   Influence of occluded air on the NPSH of a pump with a specific speed of 80 (N = 1150 r/min)

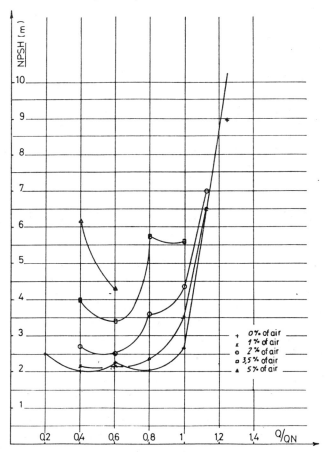

Fig 2   Influence of occluded air on the NPSH of a pump with a specific speed of 80 (N = 912 r/min)

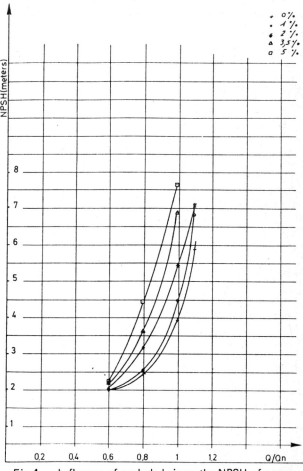

Fig 4   Influence of occluded air on the NPSH of a pump with a specific speed of 20 (N = 1500 r/min)

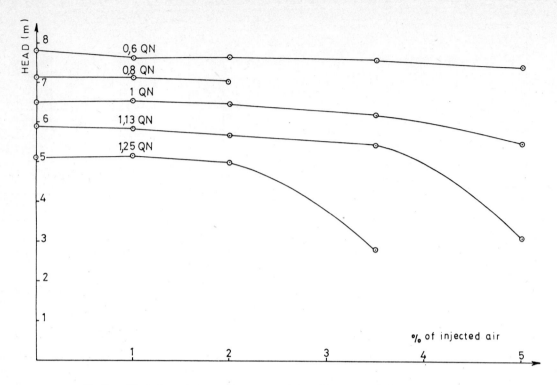

Fig 5   Variation of the total manometric head in terms of air content
          for a pump with a specific speed of 80 at 723 r/min

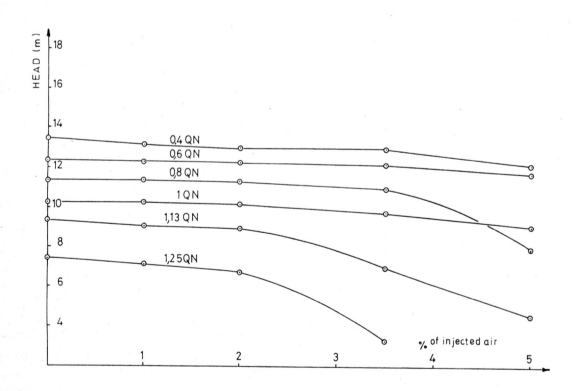

Fig 6   Variation of the total manometric head in terms of air content
          for a pump with a specific speed of 80 at 912 r/min

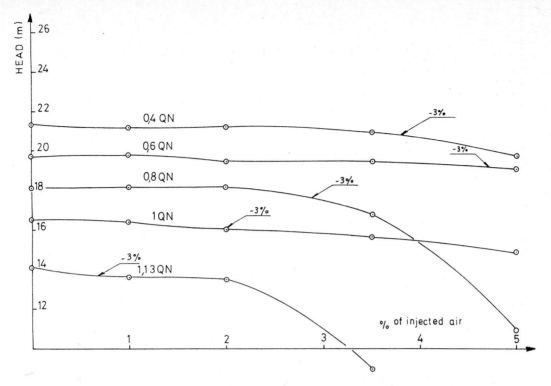

Fig 7   Variation of the total manometric head in terms of air content
for a pump with a specific speed of 80 at 1150 r/min

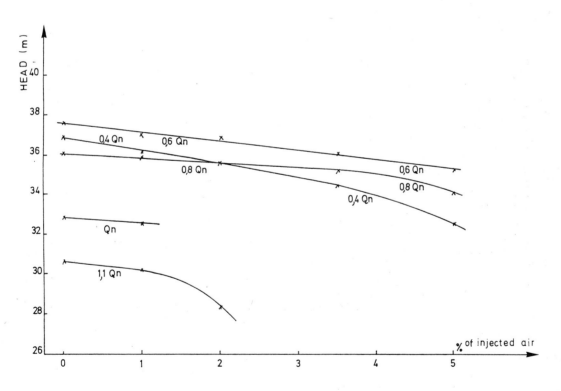

Fig 8   Variation of the total manometric head in terms of air content
for a pump with a specific speed of 20 at 1500 r/min

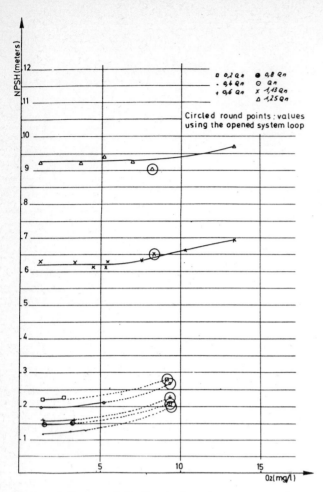

Fig 9   Influence of the dissolved oxygen content on the 3 per cent NPSH of a pump with a specific speed of 80 at 912 r/min

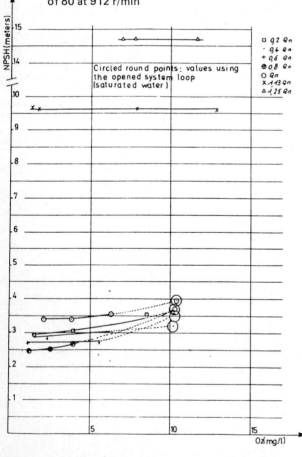

Fig 10   Influence of the dissolved oxygen content on the 3 per cent NPSH of a pump with a specific speed of 80 at 1150 r/min

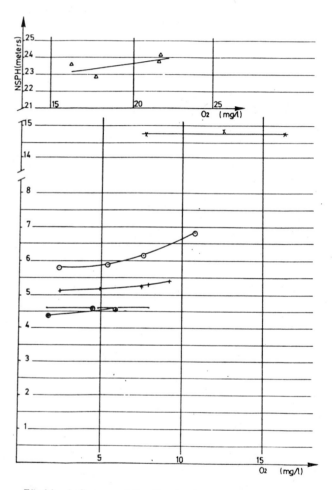

Fig 11   Influence of the dissolved oxygen content on the 3 per cent NPSH of a pump with a specific speed of 80 at 1438 r/min

C342/88

Part-load performance of inducer pumps

K OCHSNER, Dipl-IngETH
Ochsner GmBH and Company KG, Linz, Austria
O von BERTELE, BSc, CEng, MIMechE
Machinery Consultant, Norton, Cleveland

SYNOPSIS An inducer which reduces the NPSHR for centrifugal pumps from BEP down to minimum flow and stabilizes pump operation down to shut-off is described. The advantages gained by using inducers are illustrated by examples of pumps so equipped.

1 SUMMARY

An inducer is a special device fitted upstream of a conventional pump impeller to reduce the NPSH required by a pump. A limitation experienced with most inducers so far in use is the narrow flow range over which the inducer lowers the required NPSH. Outside this narrow band the suction performance of pumps so equipped is generally poorer than those without inducers. A development in Austria has resulted in an "Universal" Inducer, which reduces the required NPSH to a constant low level from about 30% to 105% of the nominal flow, by controlling the backflow in the eye of the impeller. Vibration measurements carried out both on the test bed and in the field show no increase in vibration levels with inducer equipped pumps when operated at part load, even when the fluid handled contains a high percentage of entrained gas. The advantages of inducers are illustrated with examples. Further the gradual acceleration to which fluid particles are subjected to in inducers ensures that no damage is caused to particles carried in the pumped liquid, especially when operating away from BEP. The paper concludes that the superior part load performance of pumps fitted with Universal inducers is a major contribution in the long life achieved with such pumps.

2 INTRODUCTION

Inducers-devices put in front of impellers- have long been recognised as helpful in lowering the NPSH required by pumps (1), though the limited range of their effectivness severely restricted their applications. Their development has been greatly accelerated by the need to reduce weight in machinery for space vehicles. In particular the systematic development programs of high speed pumps for liquid fuel rocket drives resulted in a rapid advance of this technology. However pumps for rockets need to operate for only a few minutes, and further only at the design rate. Pumps in the process industry and for industrial applications need to operate trouble free for years and frequently over a large range of flows. It is for the later reason that the use of inducers in industrial applications, particularly in the process industry, has been extremely limited.
In the search for maximum reliability at reasonable capital cost and a minimum of maintanance the phenomena of cavitation and its effect on the reliability of centrifugal pumps is being examined with growing interest.
On the one hand, for pumps to be reliable and to achieve high availability in the field smooth mechanical operation and long lasting seal systems must be guaranteed. Of particular importance is the need to avoid cavitation as cavitation does damage the hydraulic channels, especially those in the impeller, and so reduces the life of the pump (2). Further cavitation can also cause mechanical damage through the vibrations always associated with it.
On the other hand value analysis forces one to look for solutions which extend beyond those of past practice: the usual predicament of an engineer.
In many cases it is costly to provide the NPSH required by pumps, and frequently the size and type of pump is governed by the available NPSH. This raises the question: can the required NPSH of a pump be reduced? Changing the design, by for example using a single stage instead of a multistage pump, or by using an end suction type instead of a double suction would reduce the cost, but at the same time raises the suction specific speed (SSS). For a pump to operate satisfactorily with a high SSS it must require a low NPSH. It is for this reason that for more than 30 years attempts have been made to use inducers.

3 DEVELOPMENT OF INDUCERS.

Early designs of inducers suceeded to reduce the NPSH requirements only over a small range of flow. Some pumps were supplied with more than one inducer, so that one could change the inducer when it was planned to operate at a different rate! Further more some early designs of inducer were prone to cause damage to the pump if the available NPSH was increased, a condition always present when, for example,the level in the suction vessel changes. A development carried out in Austria resulted in an inducer design which reduces the required NPSH not only at one point but over the whole operating range of a pump Fig.1.

4 CAVITATION.

Cavitation, the formation of vapour bubbles in a liquid, is the reason that pumps fail to work if the suction pressure gets too low. To explain how an inducer helps to lower the onset of cavitation, the cavitation 'phenomena' is briefly described. When a liquid enters into the eye of an impellor it is suddenly accelerated and the increase in speed leads, according to Bernouli's law, to a drop in pressure. If the pressure at any position drops below that of the vapour pressure of the fluid then some vapour bubbles will be formed. The problem arises when the bubbles so formed collapse when the pressure of the liquid is increased. During the collapse pressures of thousands of bars can occur which will lead to damage of any solid surface nearby. Vibrations which can cause additional mechanical damage are always associated with cavitation. Furthermore in extreme cases the vapour bubbles can fill the impeller and completely interrupt the liquid flow. The required NPSH of a pump is in practice defined as that suction head where a 3% drop in the delivery head generated by the pump occurs. Dziallas (3) showed that at large flow rates cavitation occurs on the high pressure side of the impeller vanes whereas at lower rates it appears on the suction side, the two cavitation curves together producing the characteristic bathtub curve.(Fig.1)

5 INDUCERS

Inducers are a special type of impeller located in front of a convential centrifugal pump impeller. They are mounted on the same shaft. They are a kind of axial flow impeller which creates some head in front of the main impeller, and they require a lower NPSH. When an inducer is used about 7% of the total head of the stage is raised in the inducer-the head in the impeller drops by an equal amount, so that neither the overall Head/Flow characteristic nor the efficiency are greatly altered. However the initial acceleration to which the fluid is subjected when entering the inducer is small in the axial, tangential and radial direction. In the inducer itself the fluid is brought to a uniform velocity, so that by the time the fluid enters the impeller not only is it's pressure raised, but because of the uniform speed across the whole flow area, no zones of high speed with lower than average pressure are present. This combination of increased pressure and uniform velocity results in the onset of cavitation being supressed. Fig.2.

6 PART-LOAD BEHAVIOR.

To improve the part load performance of inducers measures taken to reduce the NPSHR at high rates are irrelevant- it is the second branch of the NPSH curve, the one due to cavitation on the suction side, which must be improved.
In addition to the two factors mentioned above i.e. gently increasing the pressure of the pumped liquid once it has entered the inducer, and ensuring that the velocity of the pumped fluid at the inlet of the main impellor is uniform over the whole cross-section it is essential to limit the size of the vortex cells inside the the impeller, so as to keep the recirculating liquid away from the eye of the impeller and restrict the vortex cells to a region where there already exists a raised pressure; this is mainly achieved by the uniform approach velocity established by the inducer. Also the presence of the inducer splits the region where backflow occurs: at part load

some backflow will be transferred to the inlet of the inducer where, due to the low velocity and much reduced pre rotation, it's influence on the pump's performance is greatly lowered. This shows itself in the reduced vibrations of inducer pumps at partload. Fig.3 is a record of horizontal vibrations measured on an overhang pump with (A&C) and without (B&D) inducer. Curves A and B are at shaft speed, C and D at 5 times shaft speed. It should be noted that the suction pressure was lowered when the inducer was used. At 0.11 m³/sec the suction lift was increased from 3.1 to 6.1 m, so increasing the Suction Specific Speed (SSS)*) from 3.6 to 6rps.
Fourteen parameters were found to significantly influence the performance of an inducer, among them are the length and the angle, the distance from the impeller and the surface finish of the vanes. The diameter of the inducers hub as well as the shape of the leading edge are two of the critical parameters influencing part load behavior. Further the shape of the nut securing the inducer is also significant. The advantage of pumps fitted with 'Universal' inducers, with their low NPSH requirements, together with the absence of hydraulically forced vibrations, is that they permit the process designer to use faster, smaller and often cheaper pumps, so making the Universal inducer a new tool in the continuous search for more cost effective processes.

7 GAS ENTRAINMENT

In the past Centrifugal pumps particularly in the process industry have mainly handled liquids. Nowadays there is an increasing demand, especially in biological processes, for pumps to handle multiphase fluids. Centrifugal forces in the impeller, and even more so in front of the impeller when operating away from BEP, result in a separation of gas and liquid, and consequently in a failure of the pump to produce any head. Table 1 shows the capability of different types of pumps to handle gas entrainment, it can be seen that inducer pumps perform significantly better than pumps without inducers when pumping liquids with entrained gas.

TABLE I

Type of Pump	%Gas entrained Pump can handle
Closed Impeller	5
Open "	8
Closed Impeller and Inducer	27
Half open Impeller and Inducer	35
Open Impeller and Inducer	45

This improvement is mainly due to greatly reduced prerotation i.e. the control of vortices in the suction line which are usually formed when operating at part load. In particular the capability of inducer pumps to cope with gas entrainment means that even at part load significant entrainment can be tolerated.

*) $SSS = rps \times (M^3/sec)^{1/2} / M^{.75}$

8 APPLICATIONS.

8.1 Light-weight Emergency and Fire-fighting Module

A development program sponsored by NASA for the US Navy resulted in a water pump for firefighting duties. The layout is based on the two stage fuelpumps used in spacecraft. At the design point the waterpump handles 0.32 m³/sec with a head of 106m and requires an NPSH of 2.4m, thus allowing a suction lift up to 6m. One of the requirements of the specification was that the pump had to be capable to operate at part load without any increase in NPSHR. 2.4 m corresponds to a Suction Specific Speed of 8.9 rps. The NPSHR remains at 2.4m down to 50% of the normal flowrate and the pump operates smoothly down to 30% of it's design rate. As the pump is designed to be portable (By Helicopter) it has an alluminium casing, though the shaft and the impellers are in stainless steel. The bare pump weighs only 115kg- complete with gas turbine driver, ready for use but without fuel, the weight increases to 900kg. Fig.4 shows a cross-section of the pump. Note that the driver and the second stage of the pump run at 100rps, but the speed is stepped down to 30 rps in an epicyclic gearbox to drive the first stage.

8.2 Synthesis Gas Wash Pump.

The second example illustrates how the use of a pump fitted with an inducer can considerably reduce the cost of an installation by both reducing the amount of hardware needed and also by allowing simpler elements in the construction of the pump to be used. In the production of Ammonia one of the steps is to remove Carbon Dioxide, formed during reforming, usually by washing the Synthesis gas. Pumps operating on such duties have, in the past, been frequent sources of trouble, by both suffering cavitation damage and by short bearing lives, so much so that it has been common practice to use hydrodynamic rather than rolling element bearings for such pumps. To avoid cavitation either booster pumps or slow running multistage pumps with double entry impellor have been employed. A pump installed in an Austrian fertiliser work uses a single stage single entry impeller for a gas washing duty. The pump is fitted with oil lubricated rolling element bearings and has operated for extended periods at part load. Details of the pump are shown on Fig (5).

9 GENERAL EXPERIENCE AND CONCLUSION

Over 600 inducers, with diameters between 50 and 500 mm, have been delivered by Ochsner in the last 20 years: not one had to be replaced so far (except for damage caused by a foreign body). In particular no cavitation damage has ever been observed on any inducer. In addition improved seal and bearing life is one of the features of pumps fitted with "Universal" inducers. In many instances the use of an inducer allowed a smaller, cheaper installation to be achieved by either operating a faster pump at a higher suction specific speed or by saving on the structures needed to hold vessels up in the air.

10 REFERENCES

(1) Stepanoff 2nd Ed. Centrifugal and Axial Flow Pumps.

(2) Igor J.Karassik "Pump Life under Difficult Suction Conditions"
PAE Fall Meeting Oct.1986, Irvine California.

(3) Dziallas Z. VDI 89 (1945) p.45

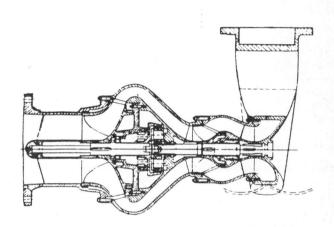

Fig 4   Light-weight emergency and fire-fighting pump
(Q = 1150 m³/h, H = 106 m, NPSH = 2.4 m)

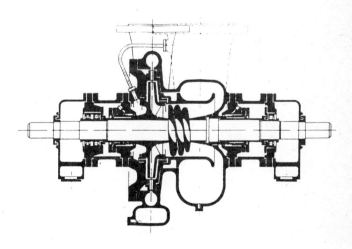

Fig 5   Pump for gas wash
(Q = 1000 m³/h, H = 300 m, NPSH = 5.5 m)

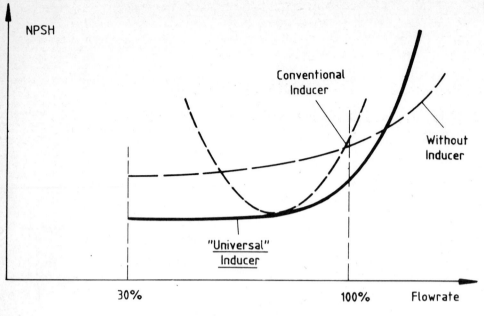

Fig 1   NPSH requirements for pumps with and without inducers

——————   Pressure Gradient with Inducer

— — —   Pressure Gradient without Inducer

—·—·—   Vapourpressure of Liquid

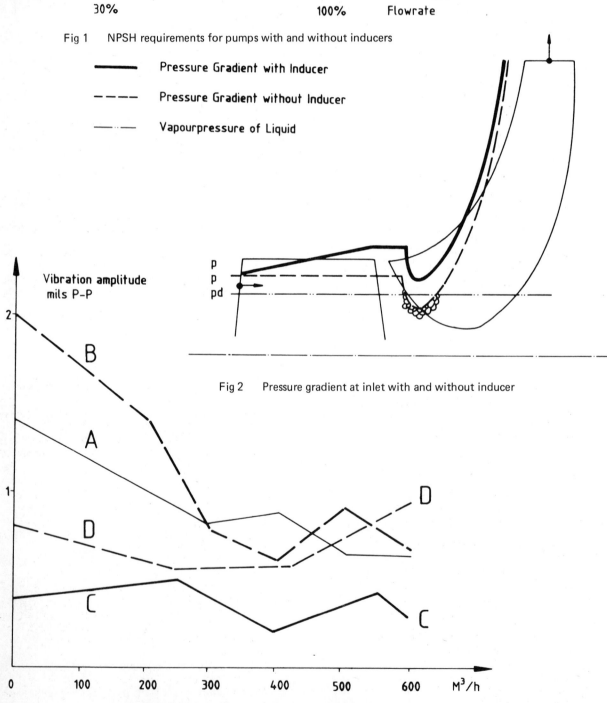

Fig 2   Pressure gradient at inlet with and without inducer

Fig 3   Vibration of pump with and without inducer

C343/88

Field problems relating to high-energy centrifugal pumps operating at part-load

L K STANMORE, DipEM, CEng, FIMechE, MASME, MBIM
Wilson-Snyder Pumps, Houston, Texas, United States of America

Part load pumping operation, investigation and analysis of actual field problems relating to horizontally split, bottom ash sluice centrifugal pumps.
This paper is concerned with an investigation and analysis of operational problems experienced with six high energy 1.12 MW centrifugal pumps operated at reduced flow and part load conditions. Methods of investigation, analysis and modifications to pump geometry to allow successful operation at reduced flow and part load are presented.

NOTATION

A	area $(mm)^2$
B	gap percentage of impeller radius
B_1	impeller water way width at inlet (mm)
B_2	impeller water way width at discharge (mm)
BM	bending moment (Kg-cm)
MR	moment of resistance (Kg-cm)
BHN	brinell hardness number
Cm_2	meridonial velocity at impeller discharge (m/s)
Cu_2	tangential velocity (m/s)
D_1	impeller eye diameter (mm)
D_2	impeller outside diameter (mm)
F_1	area between the vanes at the impeller inlet normal to the average meridonial velocity $(mm)^2$
F_2	area between the vanes at the impeller discharge normal to the average meridonial velocity $(mm)^2$
σ	stress (Kg/cm^2)
g	acceleration due to gravity (m/s^2)
H	head (m)
h_1	impeller hub diameter (mm)
h_s	impeller shroud thickness (mm)
h_v	impeller vane thickness (mm)
L	load (Kg)
l	span (mm)
n	r/min
N_s	specific speed = $\dfrac{n \times Q^{1/2}}{H^{3/4}}$
N_{ss}	suction specific speed $\dfrac{n \times Q^{1/2}}{(NPSH)^{3/4}}$
NPSH	net positive suction head (m)
P_d	discharge pressure (KPa)
P_s	suction pressure (KPa)
Pav	average pressure on vane (KPa)
Pavi	average intensity of pressure (KPa)
Pave	total average pressure on vane (KPa)
P_1 & P_2	pump case pressure at impeller eye (KPa)
Q	nominal flow (m^3/sec)
Q_{dr}	discharge recirculation (m^3/sec)
Q_{sr}	suction recirculation (m^3/sec)
Q_c	flow (m^3/sec)
R_v	resultant velocity at impeller outlet (m/sec)
R_2	impeller radius (mm)

R_3	volute radius (mm)
SF	safety factor
T_1 & T_2	thrust bearing temperature outer and inner bearings (°C)
t	impeller vane width at impeller OD (mm)
U_1	peripheral velocity at impeller eye (m/sec)
U_2	peripheral velocity at impeller discharge diameter (m/sec)
Ve	axial fluid velocity in impeller eye (m/sec)
W_1 & W_2	impeller spacing between blades at inlet and outlet - (mm)
β_1	inlet vector diagram angle (degree)
β_2	outlet vector diagram angle (degree)
p	density (Kg/m^3)
z	number of impeller vanes
b.e.p.	best efficiency point

INTRODUCTION

The application of high energy centrifugal pumps in industry requires careful matching of pump characteristics to system characteristics for reliable and economical operation. Occasionally situations develop which require the operation of existing pumps at reduced flow and part load conditons. When this occurs, unstable pump operation, mechanical failures, and system outages may follow. It is desirable to have available proven means to analyze the problems and modify the pump geometry to allow efficient and reliable operation to be restored.

A high energy centrifugal pump oversized for its intended application and operated at a greatly reduced flow and part load, provided with an inadequate by-pass piping and valves invariably will result in a high maintenance cost and the unavailability of the pump in the process system when it is most needed.

A systematic investigation was conducted in collaboration with Tennessee Valley Authority steam plant engineering staff on six bottom ash sluice water pumps. Same type and size of pump on identical service at another power plant that gave satisfactory operation over a period of sixteen years made this investigation rather challenging.

Presented here is a review of methods employed on 1.12 MW bottom ash sluice pumps operating in a fossil fuel power plant. It is expected that the methods used here will be useful in solving similar operational problems in other applications.

Description of Pumps and Installation

This paper addresses identical between bearings, double entry, single stage, double volute horizontally split case, high energy centrifugal pumps. Each pump operating singly delivers Q = .3472 m^3/sec H = 198 m, is driven by a 1.12 MW squirrel cage, drip proof, with class B insulation motors.

These pumps are installed in an existing power house to provide additional ash sluice water for bottom ash removal. Each pump takes its suction through a strainer with 3.175 mm perforation from a condensate discharge tunnel.

All six pumps are mounted in the power house basement floor elevation 132 m, taking suction from the condenser discharge tunnel centerline elevation 123 m, where the tunnel is fed from a normal river level of elevation 134 m.

These pumps originally started operation in parallel with one spare, later changed to single pump operation.

Operational History

The following problem areas and failure modes were recorded at T.V.A. during a period of seven years:
° Low frequency vibration of suction and discharge piping at reduced loads
° Instability of flow, inlet and outlet pressures
° Actual shuttling of the pump rotor assembly
° Pump shaft breakage at outboard end
° Impeller hub bore oval for 38mm length from each end
° Impeller shroud failures
° Pump case cut water erosion
° Auxiliary piping breakage - 25mm NPT
° Thrust bearing and journal bearing failures
° Cavitation noise
° High vibration amplitude at 35 HZ, i.e. within 15% of pump speed
° High noise level
° Cavitation damage in pump case inlet at the impeller eye inboard end

Numerous pump shaft and bearing failures were reported, as well as high vibration amplitudes, although rotating assemblies were balanced statically and dynamically and units were well aligned within the power plant acceptance limits.

The pumps investigated were equipped with sleeve type radial bearings and two angular contact bearings mounted back to back at the outboard end, oil ring lubricated (fig. 1).

The thrust bearings are free to move vertically so as to accommodate thrust loads only in both axial directions. Each pump operates as a single unit with the following duty requirements: Q = .3472 m^3/sec H = 198m NPSH = 4.3m at .3472m^3/sec H_2O 41.6°C n = 1785 r/min

Each unit is driven by a 1.12 Mw electric motor coupled through a gear type coupling and mounted on a beam fabricated base grouted in. The motors have a direct on line starting and reach maximum rated speed of 1785 r/min and full load in under three seconds. There is no provision between the pump and motor for a soft start.

High motor winding temperatures were reported as follows: 'A' phase, 110°C; 'B' phase, 108.8°C; 'C' phase, 112.2°C. This was thought to have caused misalignment problems due to a recorded .0889 mm growth in the motor centerline. However, this was subsequently found to be false. Motors were furnished with class 'B' insulation.

The pumps were operated over the entire flow range between shutoff and .3472 m^3/sec 847 KW with frequent starts and stops dependent on the system flow requirements. A random 'crackling' noise in the suction at the impeller eye, as contrasted with a steady 'crackling' noise associated with inadequate NPSH, could be heard 3m from the pump.

Principles and Techniques Applied in the Solution of the Problem

Pump number 5 had the worst maintenance record. Therefore, it was opened, examined and various critical observations and measurements were made (fig. 1 & 2).

Impellers were examined for dimensional compliance, casting quality, machining, balancing and any signs of cavitation or mechanical damage.

During the investigation it was discovered that the pump shafts and impellers were not obtained from the original pump manufacturer. The impellers and shaft had been manufactured with little attention to corner radii and other stress risers. Additionally, 410 stainless steel was used in place of the originally specified 17-4 pH material which has a higher wet fatigue value (275 MPa @ 100 X 10^6 cycles).

In view of the complexity of the problems experienced with these pumps, which have existed for a number of years, it was suggested that one pump be instrumented to perform a comprehensive vibration analysis in order to identify the predominent frequency of vibration and the magnitude.

Material Available for Study

° Broken shafts and damaged bearings
° Detail drawings for pump components
° Vibration analysis data on pump No.5 with and without by-pass flow
° Plant piping layouts, existing and modified
° Observing pumps in service
° Test data on pump before and after geometry changes
° Operating log sheets

Method

The first pump, No. 5, was tested on location in 1984. Four proximity probes were secured as shown in fig. 3 labeled A, B, C, and D (Table 1). Readings were taken at the

© IMechE 1988 C343/88

pump shaft sound pressure level versus frequency at the actual operating flows:
Q = .262 m³/sec -.2904 m³/sec - flow fluctuated   Pd= 105.4 KPa
P_1 = 0 to 13.79 KP_2
P_2 = 0 to 27.5KPa outboard

Thrust bearing temperatures were recorded. The vibration frequency field scan covered 0 to 500 HZ, the object was to examine important frequences as they affect a rotodynamic pump performance.

During the field test it was observed that pumps number 5, 6, and 8 operated at the following duty condition:
Pump no. 5 -
Q = .2967 - .3157 m³/sec
P_d = 1917 KPa - (2059 KPa)*
P_{1-2} = 21 KPa - 27.6 KPa
Pump no. 6 -
Q = .2390 m³/sec
P_d = 1972 KPa - (2089 KPa)*
P_{1-2} = 13.8 KPa - 27.6 KPa
Pump no. 8 -
Q = .2715 - .2811 - .3030 m³/sec
P_d=   1855 KPa (2053 KPa)*
P_{1-2} =   0 - 21 KPa

*Pressure drop most probably due to wear in the case and impeller wear rings. When the by-pass valve was open on pump no. 5, the 101.6mm by-pass pipe, together with the discharge piping exhibited low frequency vibration. Also, cavitation could be heard at the inboard bearing and suction inlet.

Table 1
Pump No. 5 - Calibration of Proximity Probes
For Location, See Figure 3.

Displacement in Mils

Probe	Static	Running	Static After Test
A	11.75	13.75	12.75
B	12.00	13.50	12.25
C	11.00	10.25	11.00
D	10.50	10.75	11.00

From the test data collected, it was found that the predominant frequency of vibration was at probe 'A' 35 HZ, i.e. 2100 r/min .226mm. With the by-pass valve open, the amplitude was .2019 mm.

Table 2
Vibration Data Pump No. 5 -

Probe	W/O By-Pass	W/By-Pass	Frequency
A	.0226mm	.2010mm	35 HZ
B	.0716mm	.0645mm	35 HZ
C	.0602mm	.0569mm	35 HZ
D	.0569mm	.0556mm	35 HZ

Pressure transducer secured to the discharge of the pump at six o'clock showed peaks at 59.16 HZ (3550 r/min) twice the speed of pump operation and 179 HZ (10740 r/min) equal to .0803 mm and with by-pass open .0902mm.

It was observed that the inner thrust bearing temperature went up to 78.3°C and the outer bearing temperature to 65.5°C from startup with an ambient of 29.4°C. Between one and two U.S. gallons per minute of cooling water was circulated through the bearing bracket water jacket.

A meter connected in the single phase of the supply did not show any power surges. It was observed that the impeller blade tips at outlet had a blunt width of 38.1mm (fig. 4).

The flow makeup for each pump is as follows: continuous requirement-.1200 m³/sec, by-pass-.0882 m³/sec, jets-.1390 m³/sec, total flow = .3472 m³/sec

Shaft displacement at probe 'B' and 'C' was recorded as +.1016 to -.1905 = .2921mm range, however, bearing examination revealed that the babbit in the bottom half was almost wiped out. This would indicate hydraulic disturbance in the volutes, caused by the blunt impeller vane tips. Even when the impeller outside diameter is the correct distance from the cut water, blunt vane tips are known to cause hydraulic hammer and disturbance in the volute [1].

Recirculation at the impeller inlet can produce pressure pulsations of 20% or more of the discharge pressure. These pulsations are the result of a phenomenon called cavitation surge. In our case, we have 1917.22 KPa X 20/100 = 383.44 KPa. This pressure pulsaton can produce an axial oscillating thrust of: 383.3 KPa$\frac{\pi}{4}(2.54^2-1.27^2)$ = 1457 Kg, a formidable destructive thrust.

Pulsating thrust will result in a rapid deterioration of the wear rings (tack welds on wear rings were broken) thrust bearing failures and pump shaft breakage. All of these failures were experienced on these pumps when operating at part load and reduced flows.

Calculations indicate that recirculation at the impeller discharge takes place at .3197 m³/sec (4937 US GPM) and at the impeller inlet at .1421 m³/sec (2250 US GPM) using W. H. Fraser Method [2].

See calculations on separate sheets.
A 13.79 KPa higher suction pressure fluctuation was observed on the outboard side of the pump. This would add another 54 Kg thrust load on the bearings. Observation of the compound pressure gauge secured at P_1 and P_2 (fig. 3) revealed positive to negative pressure changes at about .1452 m³/sec (2300 US GPM).

After completion of our test on pump No. 5, top half case was removed, various critical observations and measurements were made. Both cut water tongues were paper thin at the inlet and washed away 22.2mm in the top half case and 17.46mm in the bottom half case as measured from the pump vertical centerline.

Examination of the field pump components revealed that the shaft broke 25mm inside the impeller hub at the outboard bearings end. Maximum principle stress was calculated to be 31 MPa and the shear stress 96 MPa with a 2.5 safety factor. The tensile strength of the 17-4pH shaft material ASTM A564 type 630 is 1310 MPa and the yield 1171 MPa.

The impeller hub bore for a depth of 25mm was egg shaped at both ends of the hub. Bore showed fretting corrosion, and drawing dimensions indicated .0254 to .0762mm loose fit as standard with new impeller and shaft. The original gap 'B' on this pump prior to geometry changes as a percentage of impeller radius was less than 1%. This should be minimum 6%, maximum 12% [3].

This gap controls the noise level and the pressure pulsation which caused the damage to both cut waters. Laboratory tests with gap 'B' increased to 8.22%, showed no loss in the overall pump efficiency when the volute inlet tips were recessed.

Tack welded impeller wear rings broke loose and exhibited a rubbing pattern at 90° from horizontal pump centerline associated with hydraulic instability when operating at reduced load and flow .2210 m³/sec. Shaft failures observed did occur when the pump was operating for an extended period of time in discharge re-circulation.

Stalled areas occur at the low pressure side of the impeller vanes with rapid disappearance and reformation. Secondary flows in the impeller inlet and shroud at the outlet occur and produce damaging dynamic effects on the pump case, cut water, pump shaft and bearings [4].

Recommendations

On the basis of the various observations and calculations made and the data collected, it was recommended that the following corrective steps be taken.

° Increase by-pass piping from 102mm to 152mm so that the pump will not operate at or below .3157 m³/sec (5000 U.S.GPM) in any system mode

° Increase gap 'B' from 1% to give $B = 100 \dfrac{(362 - 328)}{362} = 9.4\%$ by reducing the impeller outside diameter to 657.2 mm diameter and underfiling the vane tips at outlet to obtain the original head 216.4m (710 ft) at .3472 m³/sec (5500 US GPM) and to reduce the hydraulic disturbance. Gap 'B' controls the strength and amplitude of hydraulic shock created by vane passing frequency.

° Impeller shrouds must be equal distance from pump case wall to minimize hydraulic axial forces.

° Reduce gap 'A' to 1.27mm (fig. 5) by welding case reducing rings. Gap 'A' controls the severity of pressure pulsation behind the impeller shroud giving rise to high axial directional forces.

° Use thrust bearings with 40° contact angle for improved thrust load carrying capacity.

° Impeller shroud at outside diameter to be full width 7.93mm (fig. 5) to achieve a better impeller to pump case seal.

Phase Two of Investigation

The operator's engineers and staff accepted the recommendations and pumps no. 6 and 7 were modified and tested at the pump manufacturer's test facility. The case sealing ring was fabricated from a steel strip rolled, welded, and machined in position. No preheat or post-heat was required. The pump case was subjected to a hydrostatic test to insure integrity of casting in the machined areas receiving the sealing rings.

Pump was installed in the field and instrumented as shown in fig. 3. Six proximity probes were used to monitor vibration. Compound pressure gauges P_1 and P_2 were mounted on each side of the pump.

Panel mounted pressure gauges, a digital flow meter, together with T_1 and T_2 thermocouples secured on top of the thrust bearing bracket were used in this test. Readings were taken and recorded at various flows and pressures, together with frequency and amplitudes (fig. 6, tables 3, 4, 5, and 6).

Vibration frequency field was 0 to 500 HZ (fig. 6 & 7). The object was to examine important frequencies as they affect the pump performance, in particular the 35 HZ, which provided the basis for comparing the test data collected before the modifications to pump no. 5 and compare with test data and results collected on pumps no. 6 & 7 after incorporating the geometry changes recommended in phase one of this investigation.

No. 6 pump, fitted with 676.3mm diameter impeller and incorporating the new geometry changes, was run continuously for 208 days; then it was taken out of service to accommodate pumps No. 6 and No. 7, both fitted with 654mm impellers.

Pump top case was removed and internal parts examined. No rubbing on the wear rings or sealing rings was observed. Machining tool marks were clearly visible on the impeller periphery.

Results After Geometry Changes

Thrust bearing temperature at .2904 m³/sec was 78.3°C during the 1984 test. With the new geometry, this was reduced to 36.6°C and 37.7°C respectively on pumps no. 6 and 7. The original high bearing temperature can be traced to probe no. 6-A.

The predominant frequency of vibration at probe no. 6 was originally 35 HZ, which is within 14.29% of the synchronous speed of the pump operation, with an amplitude of .226mm and .202mm. New test data shows this to be .0025mm and .0076mm range. This value represents the shuttling hydraulic force and was responsible in great measure for the numerous thrust bearing failures.

The 1984 test indicated random suction pressure fluctuation of 0-13.79 KPa at point P_1 and 0-27.58 KPa at point P_2, a formidable axial oscillating thrust of:

$= P_s \dfrac{\pi}{4} (D_1{}^2 - h_1{}^2)$

$= 27.58 \text{ KPa} \dfrac{\pi}{4} (2.54^2 - 1.27^2)$

$= 104.8 \text{ Kg.}$

This is in addition to the recirculation at the impeller inlet which can produce pressure pulsation, as stated previously, in the order of 20% or more of the pump discharge pressure generated. These unstable low flow recirculation and pressure pulsations at reduced load have been greatly reduced as indicated by the uniform non-fluctuating pressure observed on both sides of the impeller suction inlet $P_1 = P_2$ at tested flow of Q = .4104, .3598, .3157, .2967, .2588, .2399 m³/sec.

Pumps run smoothly with no audible recirculation noise or cavitation. The 1984 test exhibited wide pressure gauge fluctuation in the .3157 to .2399 m³/sec flow range.

The motor end bearing on pump no. 6 at 30 HZ indicated .0813mm value in the horizontal plane, however, the motor was found to be out of balance; also, one short bolt in the gear coupling was located. Correcting both areas rectified this problem.

A .0813mm value translates into the rough zone of operation and did affect the pump bearing readings at the coupling end probes no. 3 and 5.

There was no visible damage on pump No. 6, which incorporated the geometry changes and ran successfully for 208 days.

Test data indicated that the point of discharge flow re-circulation has been moved to the left of .3157 m³/sec., which was the objective (fig. 8). Furthermore, the high .2261mm peak to peak value at probe 'A', now probe 6 has been reduced to .0025 to .0076mm range. The pump case and impeller geometry changes recommended in 1985 and implemented on pumps No. 6 and 7 logged 208 days of trouble free operation have proven to be a notable improvement and indicate one suitable approach for retrofitting existing oversized pumps, which have to operate at reduced part load. Customer reported the following: no shaft breakage, no thrust bearings lost, no secondary piping broken, and thrust bearings temperature dropped from 78.3°C to 37.7°C.

Impeller Shroud Failure

After some six months of continuous operation at partial load and below the recommended minimum flow of .3157 m³/sec (5000 US GPM), the impeller in pump No. 6 was found to have shroud broken. Examination of the defective impeller revealed that the impeller was typical production casting 13-4 stainless steel ASTM A296 grade CA6-NM.

The impeller shroud broke off 229 X 76mm portion at the outside diameter in one location only. This failure was identified as a fatigue cracking that initiated from the inside of the impeller at the radius between the impeller vane on the shroud. The crack propagated toward outside of shroud and the impeller outside diameter where it finally broke off. The broken off portion of the shroud was not available for metallurgical study.

The outside impeller shroud showed signs of grinding and polishing, typical results of impeller static and dynamic balancing. Machining tool marks were clearly visible at the impeller shroud outside diameter indicating that there was no rubbing contact between the outside diameter of the impeller and the pump case sealing rings. Dye penetrant examination did not reveal any significant indication at the fracture surfaces and the surrounding areas.

Hardness was 286-310 BHN, which appears to be on the high side for a CA6 NM stainless steel casting. The casting has radius of about 3.6m between each vane and the shroud and the impeller was found to be within the design tolerances.

The broken portion of the shroud was analyzed for normal mode of loading under dynamic conditions and the material strength. It was assumed that twice the average pressure due to change of momentum exists at the outer end of the vane. This average pressure is acting on the most outer unit strip at the shroud between the vanes. Analysis of the average pressure acting on the shroud under normal operating conditions indicate a 5.6 application factor for the material, giving ample strength. The impeller analyzed represents a typical normal sand casting with no deficiency sighted in the material.

It is evident that the missing shroud portion was subjected to abnormal internal stresses. This is a result of operating the pump at partial load and in the discharge recirculation mode for a prolonged period of time. When the pump is operating in an impeller discharge re-circulation mode (calculations indicate this to be occurring at .312 m³/sec [4937 US GPM]), a portion of the flow is getting back into the impeller channels and cavitation is caused by local vortices locked into the space between the vanes.

The pressure pulsation produced by cavitating vortices on the vane surface between the shrouds is of sufficient magnitude and frequency that the impeller shroud fails by fatigue, which was observed in this case.

See calculations in appendix

Conclusions

It would appear that a minimum safe value for recirculation on these pumps is 45% of b.e.p. at maximum impeller diameter. Any lower value will result in major problems. It was further observed that by-pass piping, when provided, requires careful sizing and valving to insure that the estimated minimum flow is not reduced any further. Existing installation had a 100mm pipe, but only a 76mm pneumatically operated valve.

Horizontally split pumps with double entry impellers appear to be very sensitive to piping arrangement, both in the suction and discharge. Strainer boxes on the pump inlet should be well sized. Discharge piping requires attention, not only for hydraulic throughput, velocity, friction losses, and piping supports; but provision must be made for suitable anchoring of pipes to accommodate reaction forces developed by flow and pressure in the discharge piping. In our case, 36m long when pipes have a 90° elbow. It was observed that, at the reduced flow and part load, the 356mm discharge piping was moving laterally about 12.7mm and deflecting in the vertical direction between the pipe supports 25 to 38mm at the discharge recirculation capacity of .2525 m³/sec prior to the geometry changes.

A double volute design is essential in this type of application and can greatly reduce the radial force, which is a resultant vectorial value. A single volute pump is not suitable for this type of application due to the increasing radial force at the reduced flow and load reaching 100% at shut off.

The pumps evaluated have relatively large impeller eye areas, in our case, the difference in suction pressure between the two inlets of only 27.6 KPa produced an axial oscillating thrust of 104.8 Kg. For this very reason, double suction, double entry impeller pumps require thrust bearing arrangement of ample capacity, a pivoted segmental-type thrust bearing would be in order.

It is suggested that the impeller central rib extend to the impeller periphery. More vanes will increase the shroud resistance to moment. Also, vanes should be staggered to reduce vibration amplitudes at vane passing frequencies [5]. Those modifications were not incorporated in our case. The number of impeller vanes and the outlet vane angle has a big effect on the head rise characteristics. Fewer vanes, in our case-five, do give more pressure pulsations with a lower frequency. The pump 'A' and 'B' dimensions, impeller vane angle, eye diameter must be well matched to the duty requirements, as all parameters influence the pump hydraulic stability.

Axial displacement or 'shuttling' was observed at proximity probe 6 (value recorded .226mm) at partial load and reduced flow. After geometry changes, this was reduced to .0037mm. This axial shaft movement was of random nature and is associated with discharge recirculation. The shuttling phenomenon at partial load and reduced flow requires further study for better understanding.

This investigation has demonstrated that the geometry changes implemented on the high energy ash sluice pumps did allow the oversized pumps to operate at reduced flow and partial load. A 45% re-circulation flow can be maintained for satisfactory operation.

It was further demonstrated that gap 'A' is a very important system vibration parameter. Shaft and bearing failures were eliminated as a result of the geometry changes to gap 'A'.

The low frequency vibration amplitudes at 35 HZ .226mm were reduced to .0037mm demonstrating the controlling effect gap 'A' has on the pressure pulsation behind the impeller shroud which did give rise to hydraulic axial forces.

Another way of addressing this type of problem would be to design a low flow impeller, or replacing the oversized pumps with smaller capacity units. In the latter case, one has to consider the capital investment involved in selecting a smaller, better fitting pump, changes in piping, foundation, driver starting equipment, interchangeability of components with existing pumps, and many other criteria.

Acknowledgement

This work has developed from actual field problems on pumps operating at part load and reduced flows. Test data was obtained at Wilson-Snyder Plant, McAlester, Oklahoma and in collaboration with Tennessee Valley Authority at one of their fossil fuel power plants where these pumps are located.

The author acknowledges with thanks the assistance of T.V.A. engineering staff and the use of their facilities and instrumentation in this investigation.

The author would like to thank National-Oilwell company for permission to publish this paper.

References

1. HEALD, C. C. and PALGRAVE, R. Backflow Control Improves Pump Performance, Ingersoll-Rand, Phillipsburg, N.J., reprinted 1985 Oil and Gas Journal, U.S.A.

2. FRAZER, W. H. Recirculation in Centrifugal Pumps, Winter Annual Meeting of ASME, 1981, Washington, D.C., McGraw Edison Worthington, U.S.A.

3. NELSON, W. ED Maintenance and Trouble Shooting of Single Stage Pumps, First International Pump Symposium Proceedings Houston, Texas, U.S.A.

4. MAKAY, ELEMER Hydraulic Problems in Nuclear Feed Pumps Demand Attention to Cut Costly Down Time, 1975, Vol. 119, No. 3 Power, U.S.A.

5. MAKAY, E. and O. SZAMODY Recommended Design Guidelines for Feedwater Pumps in Large Power Generating Units, CS-1512 Research Project 1266-18 final report 1980, Energy Research and Consultants Corporation, 900 Overton Avenue, Morrisville, PA 19067

Table 3

Readings Recorded After Geometry Changes
Pump No. 6   654 mm Diameter Impeller

Q = .4104 m^3/sec	.3157-.2967	.3472-.3598	.3157	.2652-.2399	.2399-.2588
P_d = 1827.6 KPa	2027.5	1965.5	2000	2096.5	2096.5
P_s = 3.44 KPa	3.44	3.44	3.44	6.89	6.89
P_1 = 10.34 KPa	10.34	10.34	10.34	10.34	10.34
P_2 = 13.79 KPa	13.79	13.79	13.79	20.69	20.69
T_1 = 25.5°C	26.6	35	36.6	37.7	37.7
T_2 = 25.5°C	26.6	35	36.6	37.7	37.7

Table 4

Proximity Probe Voltage Levels - Pump No. 6

Conditions: Initially set to 0, conversion factor 200 mv/.0254; positive value indicates movement toward probe.

	at .2525 m^3/sec flow			at .3157 m^3/sec flow	
Probe	Voltage/Volts	Displacement/Mils	Probe	Voltage/Volts	Displacement/Mils
1	+ .105	+ .127	1	+ .120	+ .152
2	+ .165	+ .203	2	+ .185	+ .228
3	+ .900	+ 1.143	3	+ .770	+ .990
4	− .375	− .482	4	− .270	− .355
5	− .162	− .203	5	− .135	− .177
6	− .040	− .050	6	− .015	− .025

Table 5

Readings Recorded After Geometry Changes
Pump No. 7   654 mm Diameter Impeller

*Q = .2652-.2967 m^3/sec	.2652-.2967	.322-.3598	.322-.3598	.322-.3598
P_d = 2068.9 KPa	2068.9	1986	1986	1986
P_s = 6.9 KPa	10.34	0	0	0
P_1 = 13.79 KPa	13.79	13.79	13.79	13.79
P_2 = 20.69 KPa	20.69	13.79	13.79	13.79
T_1 = 28.3°C	37.7	37.7	37.7	36.6
T_2 = 28.3°C	28.3	35	33.8	33.8
Time: 1:50 PM	2:05 PM	2:15 PM	2:30 PM	2:45 PM

* These flows are system dependent and are set automatically depending on the number of nozzles open.

Table 6

Proximity Probe Voltage Levels Pump No. 7

Conditions: Initially set to 0, conversion factor 200 mv/mil; positive value indicates movement toward probe.

	at .2841 m^3/sec flow			at .3346 m^3/sec flow	
Probe	Voltage/Volts	Displacement/Mils	Probe	Voltage/Volts	Displacement/Mils
1	--------Probe Jarred-------		1	+ 2.85	+ .355
2	+ .260	+ .833	2	+ .290	+ .036
3	+ .652	+ .083	3	+ .768	+ .096
4	− .390	− .049	4	− .245	− .030
5	+ .561	+ .071	5	+ .565	+ .071
6	− .010	− .001	6	− .050	− .007

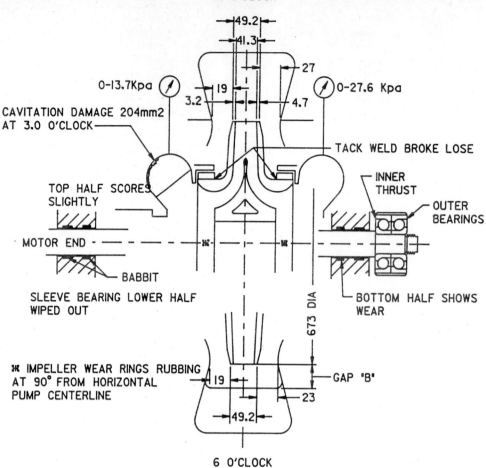

Fig 1   Plan view of pump case with top half removed; test pump
         number 5 before geometry changes

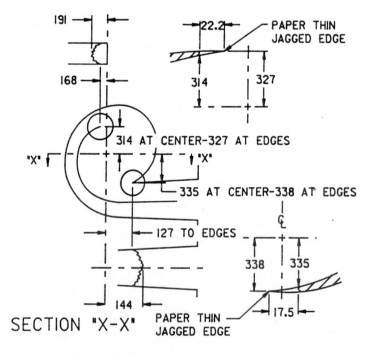

Fig 2   Test pump number 5 before geometry changes

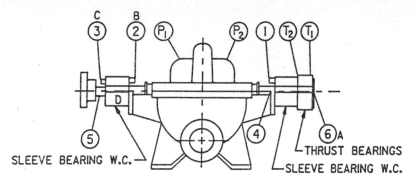

I, 2, 3, 4, 5 & 6 POSITION OF PROXIMITY PROBES.

T_1 & T_2 THERMOCOUPLES - THRUST BEARING.

P_1 & P_2 COMPOUND SUCTION GAUGES.

Fig 3   Test instrumentation for ash sluice pump

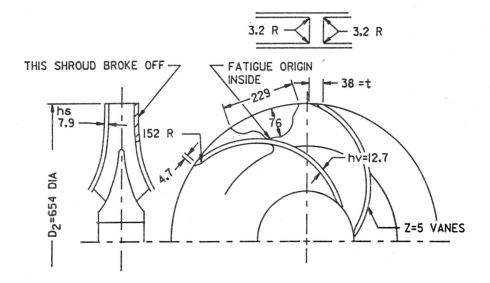

Fig 4   Pump number 6 impeller

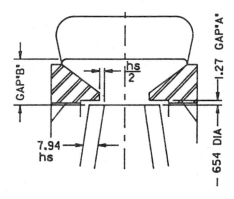

Fig 5   Geometry changes to ash sluice pumps

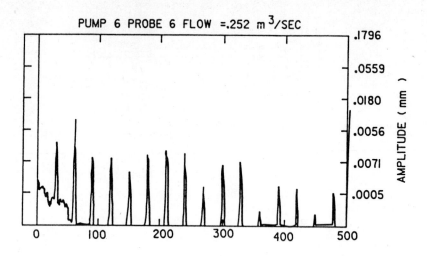

Fig 6 Vibration recorded on pump number 6 after geometry changes

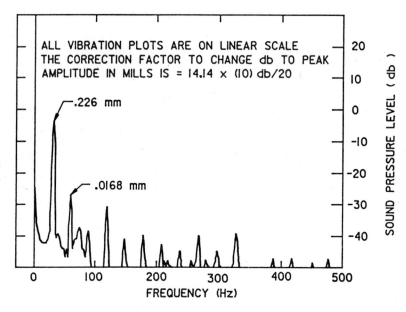

Fig 7 Vibration recorded on pump number 6 before geometry changes

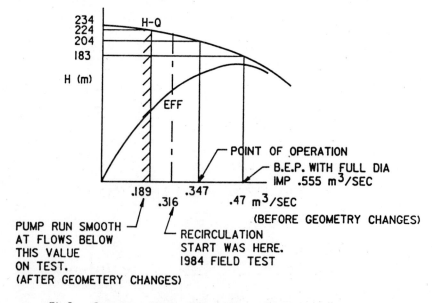

Fig 8 Geometry changes effect on pump performance

APPENDIX

RECIRCULATION CALCULATION

B_2 = 55.56 mm $\quad\quad$ Q = .555 m /SEC

B_1 = 41.27 mm $\quad\quad$ H = 173.7 m

h_1 = 136.52 mm $\quad$ N = 1750 r/mm

D_1 = 254 mm $\quad\quad$ NPSH = 7.3 m

W_1 = 54.76 mm

W_2 = 101.6 mm

D_2 = 673.1 mm

$$F_2 = B_2 \times W_2 \times \text{No VANES}$$
$$= 41.27 \times 101.6 \times 5$$
$$= 20965.16 \text{ mm}^2$$
$$F_1 = B_1 \times W \times \text{No VANES} \times 2$$
$$= 55.56 \times 54.76 \times 10$$
$$= 30425 \text{ mm}^2$$

$$N_S = \frac{h \times Q^{1/2}}{H^{3/4}} = \frac{1750 \times .5555^{1/2}}{173.7^{3/4}} = 27.25$$

$$N_{SS} = \frac{h \times Q^{1/2}}{NPSH^{3/4}} = \frac{1750 \times .5555^{1/2}}{7.3^{3/4}} = 293.6$$

$$U_1 = \frac{\pi D_1 n}{60} = \frac{\pi \times 254 \times 1750}{60} = 23.27 \text{ m/SEC}$$

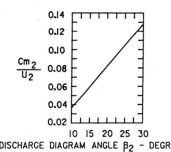

DISCHARGE DIAGRAM ANGLE β_2 – DEGREES

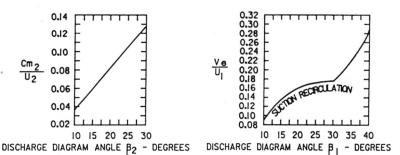

DISCHARGE DIAGRAM ANGLE β_1 – DEGREES

SUCTION RECIRCULATION

$$U_2 = \frac{\pi D \times n}{60} =$$
$$= \frac{\pi \times 673 \times 1750}{60} = 61.66 \text{ m/SEC}$$

DISCHARGE RICIRCULATION

$$S \text{ In } \beta_2 = \frac{F_2}{\pi \times D_2 \times B_2} = \frac{20965.16}{\pi \times 673.1 \times 41.27} = .2403$$

$$S \text{ In}^{-1} \quad .2403 \quad 13.9$$

$$\frac{Cm_2}{U_2} = .058 \text{ FROM FIG 10}$$

$$FLOW = \frac{D_2^2 \times B_2 \times r/mm}{6.089 \times 10^9} \times \frac{Cm_2}{U_2} = \frac{673.1^2 \times 41.27 \times 1750}{6.089 \times 10^9} = .058$$

$$= .3117 \text{ m}^3/\text{SEC} \text{ (4937 US GPM)}$$

SUCTION RECIRCULATION

$$S \text{ In } \beta_1 = \frac{1.273 F_1}{2 (D_1^2 - h_1^2)} = \frac{1.273 \times 30424.65}{2 (254^2 - 136.52^2)} = .4221$$

$$S \text{ In}^{-1} \quad .4221 \quad 24.96$$

$$\frac{Ve}{U_1} = .167 \text{ FROM FIG 11}$$

$$FLOW = \frac{D_1 (D_1^2 - h_1^2) r/mm}{2.43 \times 10^{10}} \times \frac{Ve}{U_1} = \frac{254 (254^2 - 136.52^2) 1750}{2.43 \times 10^{10}} \times .167$$

$$= .1401 \text{ m}^3/\text{SEC} \text{ (2219.2 US GPM)}$$

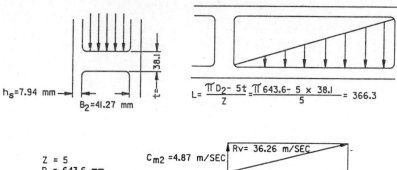

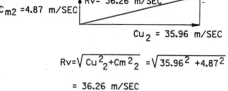

$$Z = 5$$
$$D_2 = 643.6 \text{ mm}$$
$$Qc = .3472 \text{ m}^3/\text{SEC}$$
$$H = 198 \text{ m}$$
$$n = 1750 \text{ r/min}$$
$$\ell = 999.6 \text{ Kg/m}^3$$
$$g = 9.82 \text{ m/SEC}^2$$

$$L = \frac{\pi D_2 - 5t}{Z} = \frac{\pi\, 643.6 - 5 \times 38.1}{5} = 366.3$$

$$C_{m2} = 4.87 \text{ m/SEC}$$

$$Rv = 36.26 \text{ m/SEC}$$

$$Cu_2 = 35.96 \text{ m/SEC}$$

$$Rv = \sqrt{Cu_2^2 + Cm_2^2} = \sqrt{35.96^2 + 4.87^2}$$

$$= 36.26 \text{ m/SEC}$$

CHANGE OF MOMENTUM PER SECOND

$$= \frac{Qc \times \ell \times Rv}{g} = \frac{.3472 \times 999.6 \times 36.26}{9.82} = 1282.2 \text{ Kg}$$

WE HAVE 5 VANES EACH 685.8 mm LONG × 35 mm WIDE AVERAGE

TOTAL VANE AREA
$$= 5 \times 685.8 \text{ mm} \times 35 \text{ mm}$$
$$= 1200.15 \text{ mm}^2$$

AVERAGE INTENSITY OF PRESSURE

$$Pavl = \frac{1}{A} = \frac{1282.2 \text{ Kg}}{1200 \text{ cm}^2} = 1.07 \text{ Kg/cm}$$

TWICE THE AVERAGE PRESSURE MAY BE ASSUMED TO EXIST AT THE OUTER END OF VANE :

$$Pav = Pavl \times 2$$
$$= 1.07 \text{ Kg/cm}^2 \times 2$$
$$= 2.14 \text{ Kg/cm}^2$$

CONSIDER THE OUTER MOST 1 cm OF VANE AS A BEAM FIXED AT ITS ENDS WHICH ARE 41.27 mm APART

$$BM = WL/12$$
$$= \frac{2.14 \text{ Kg/cm}^2 \times 4.127 \text{ cm} \times 4.127 \text{ cm}}{12}$$
$$= 3.04 \text{ Kg} - \text{cm}$$

MOMENT OF RESISTANCE IF THICKNESS OF VANE BE DENOTED BY h_v

$$MR = \frac{\sigma\, h_v^2}{6}$$

$$3.04 \text{ Kg} - \text{cm} = \frac{703 \text{ Kg/cm}^2 \times h_v^2}{6}$$

$$h_v \sqrt{\frac{3.04 \text{ Kg} - \text{cm} \times 6}{703 \text{ Kg/cm}^2}}$$

$$= .161 \text{ cm}$$
$$= 1.61 \text{ mm}$$

OUR VANE THICKNESS IS 12.7 mm

IMPELLER SHROUD STRENGTH

CONSIDER ONE UNIT STRIP ACTING AS A BEAM OF L=336.36 mm
SPAN, THE PRESSURE AT ONE END MAY BE TAKEN TO EQUAL
TO THAT ON THE EXTREMITY OF VANE vIz 2.14 Kg/cm^2
AND 'O' AT THE OTHER END.

TOTAL PRESSURE ON SHROUD :

$$Pav_t = \frac{2.14 \ Kg/cm^2 \times 36.63 \ cm}{2} \times PER/cm$$

$$= 39.2 \ Kg/cm^2$$

BM DUE TO THIS VARYING LOAD IS NOT GREATER THAN
THAT DUE TO 39.2 Kg/cm   UNIFORMLY DISTRIBUTED
I.e. NOT GREATER THAN :

$$BM = WL/12$$

$$= \frac{39.2 \ Kg/cm^2 \times 36.63 \ cm}{12}$$

$$= 119.7 \ Kg - cm$$

MOMENT OF RESISTANCE IF SHROUD IS = h_s = 7.937 mm

$$MR = \frac{\frac{\sigma \ bh_s^2}{12}}{h_s/2} = \frac{\sigma \ bh^2}{6}$$

$$= \frac{703 \ Kg/cm^2 \times 36.63 \ cm \times (.396)^2}{6}$$

$$= 673 \ Kg - cm$$

$$SF = \frac{MR}{BM} = \frac{673}{119.7} = 5.62$$

σ = 703 Kg/cm^2
ALLOWED FOR THE STRAINING ACTION
PRODUCING EQUAL STRESS OF OPPOSITE
SIGN ALTERNATIVELY

SHROUD OF 7.94 mm IS STRONG ENOUGH IN CA6NM CAST MATERIAL

C344/88

Part-load operation of the boiler feedwater pumps for the new French PWR 1400 MW nuclear plants—a challenge for the designer

R MARTIN
Électricite de France, Villeurbanne, France
A VERRY
Électricite de France, Chatou, France
J POULAIN and **G VERDONK**, PhD
Alsthom-Rateau, La Courneuve, France
R CANAVELIS, PhD and **P GUILLOISEAU**
Bergeron SA, Paris, France

SYNOPSIS

The Boiler feedwater pumps in EDF Power Stations have to work reliably for all flow rates between 33 % and 134 % of the flow corresponding to the best efficiency point (Q_{BEP}). Under transient conditions associated with load changing these limits increase to 33 % to 147 % Q_{BEP}.

Due to the high specific power of these pumps, the operating conditions heavily influence the hydraulic and mechanical dimensioning. This paper presents some particular aspects of their design and test results obtained on a pump model as well as on a full scale prototype concerning suction performance, head capacity curve stability, pressure pulsations and structural excitations.

1 - INTRODUCTION

ELECTRICITE DE FRANCE (E.D.F.) has gained an extensive experience in operating large boiler feed pumps in about 50 nuclear power plants now in service. These plants have been designed by series of different types of units called successively 900 MW CP1 (24 units), 900 MW CP2 (10 units), 1300 MW (12 units). Owing to this large field of experience, the actual operating conditions of power plants components have been better defined and, consequently, the performance of the new equipment is expected to be improved. More particularly the part load operating conditions of boiler feed pumps have been highlighted and studied so that the resulting experience can be used to define the specifications of those pumps in new 1400 MW "N4" units. Since 70's, E.D.F has been providing two turbine driven half capacity boiler feed pumps (B.F.P) into each nuclear PWR type power unit. Therefore, in case of a trip of one of the two B.F.P., E.D.F requires two automatic control operations :

1/ The output load has to be quickly decreased from 100 % to about 65 % of the nominal load.

2/ The flowrate of the BFP which is still in operation has to be increased from 100 % to 130 % of its nominal capacity. From these requirements it can be seen in figure 1 that the BFP has to operate in a large flow range, from about 33 % to 147 % of QBEP.
Consequently, matching the hydraulic design to overload capacities makes the part load operation even more difficult.

Taking these remarks into account in order to obtain the maximum availability especially at part load, E.D.F. has required the following specifications :

a/ an experienced and well known technology : main pump and booster pump with a double inlet impeller, volute and diffuser for main pump and double volute for booster pump, barrel type casted casing, two forced oil lubricated bearings, one double thrust bearing, mechanical seals.

b/ no bubble in the main pump impeller eye for all loads within the normal flow range (see figure 1). This "bubble free" cavitation criterion may appear slightly conservative. However it gives the user an unquestionable warranty about impeller life. As a matter of fact, E.D.F prefers ensuring a long life duration to the pumps by specifying such a cavitation criterion rather than defining a given number of safe operating hours.

c/ a continuously decreasing head capacity curve in order to make parallel as well as single pump operation stable despite the very flat system characteric curve.

d/ the hydraulic performance had to be checked first on a 0.73 scale model and then on a full scale and industrial made prototype. Although the tests are done at low speed they are supposed to be sufficiently representative of the actual operating conditions to avoid any major trouble during the site commissionning.

e/ as low as possible pressure pulsation level and a as good as possible prediction of pump/circuit interactions. All pumps create hydraulic forces which can produce vibrations of bedplates, basement, and piping, particularly at part load. These broad frequency band forces will obviously cause unwanted responses of the surroundings of the pumps corresponding to their natural frequencies.

2 - PUMP SUCTION PERFORMANCE

The hydraulic design of the impeller has been developed on the basis of an initial design previously used in boiler feed pumps for 900 MW power units and for a proposed 1300 MW plant BFP.

The required NPSH curves corresponding to the initial design are represented with dotted lines in Figure 2. The three different curves which are shown give the following NPSH values :

$(NPSH)_F$ corresponding to incipient cavitation (no bubble)

$(NPSH)_d$ corresponding to 0 % head drop.

$(NPSH)_c$ corresponding to 3 % head drop.

It should be noted that, as often seen in many normal impeller designs, these dotted line curves show a peak for a critical flow corresponding to the onset of recirculation at impeller inlet. At partial flows lower than this critical flow, unsteady cavitation sheets are located on the leading edge of impeller blades. It has been proved that operating a pump in such conditions for extended periods may lead to severe cavitation damage.

It is the reason why, in order to prevent any cavitation damage in the whole operating range, the shape of the inlet portion of impeller blades has been successively adapted by means of visual model tests. The adaptation consists essentially in a special streamlining which leads to a matching of inlet flow angles in a wider flow range. The resulting NPSH curves corresponding to this new design are shown in figure 2 with unbroken lines. The new required NPSH values are slightly higher at nominal flow but substantially smaller at partial flow. As a result the available NPSH at pump inlet is alway higher than the required $(NPSH)_F$ corresponding to no bubble at impeller blades inlet, with a substantial NPSH margin in the whole operating range.

The NPSH performance and cavitation visualisation tests have been performed with a 0.73 scale model and with a specially adapted full scale prototype. Cavitation bubbles were observed with stroboscopic light by means of adequately located windows which allowed to visualize both suction and pressure sides of blades. The test results have been found quite similar in model and prototype with discrepancies lower than measurement uncertainties.

The prototype has been tested at two different rotational speeds : 1000 rpm and 1500 rpm. The cavitation test results were also the same in both conditions.

3 - HYDRAULIC STABILITY AT PARTIAL FLOW

In order to prevent pump and system hydraulic instabilities, the pump has to produce a continuously increasing head with decreasing flow.

A stable head-flow curve can be obtained with a proper hydraulic design of suction casing, impeller and diffuser. Among the many influential factors (1), (2), the following ones may be regarded as having a major effect on the head-flow curve at partial flow :

- $b3/b2$: diffuser inlet b3 width over impeller outlet width ratio b2.

- $D3/D2$: diffuser vanes inlet diameter D3 over impeller blades outlet diameter D2 ratio (gap B of reference (1)).

- $D'3/D'2$: casing internal diameter D'3 over impeller peripheral diameter D'2 ratio (gap A of reference (1))

- Cylindrical or conical cutting of impeller blades outlet edge.

- Extension of impeller shrouds.

- Diffuser channels inlet area and shape.

These factors have been successively modified during model and prototype tests in order to find a fully satisfactory head-flow curve. Nevertheless, D3/D2 ratio had to be maintained high enough in order to keep impeller-diffuser interactions as low as possible. Some of the results of the development program are reported below.

3.1 First series of tests

Figure 3a shows several head-flow curves obtained in a first series of model tests. Curve A corresponds to a basic test performed with the initial hydraulic design. The head-flow curve appears to be stable with the following values of geometrical factors mentioned above :

$b3/b2 = 1.08$ $D3/D2 = 1.06$ $D'3/D'2 = 1.017$

Curve B corresponds to a higher diffuser inlet diameter. This modification was made in order to adapt the best efficiency point to higher flows. Consequently, the ratio D3/D2 and the diffuser channel inlet area both increased :

$b3/b2 = 1.08$ $D3/D2 = 1.07$ $D'3/D'2 = 1.017$

The head-flow curve became unstable with a "droopy" portion at low flow.

With the same ratios $b3/b2 = 1.08$ and $D3/D2 = 1.07$.

Curve C has been obtained by reducing the ratio D'3/D'2 to 1.008. It may be noted that this modification proved to be sufficient to produce a stable curve.

3.2 Second series of tests

Figure 3b gives several head-flow curves obtained in another series of model tests intended to check the performance of a double suction impeller with staggered blades.

Previous head-flow curve C in figure 3a corresponded to an impeller with aligned blades. The new impeller was tested in the same environment (suction casing and diffuser).

Fig. 3b curve A corresponds to an impeller outlet diameter slightly higher than that of fig. 3a curve C :

b3/b2 = 1.08   D3/D2 = 1.05   D'3/D'2 = 1.008

Despite the lower value of the ratio D3/D2, curve A proved to be unstable.

Curve B has been obtained with a conical cutting of impeller blades outlet edges. The ratio D3/D2 was unchanged near central shroud and increased to 1.08 near the side shrouds. The ration D'3/D'2 remained unchanged. Decreasing the average impeller outlet diameter made the whole head-flow curve shift downwards. It can be seen that despite the higher value of radial gap D3/D2, head-flow curve B appears to be less unstable than curve A.

Curve C is the result of a further change in impeller geometry. The ratio D3/D2 near side shrouds was increased to 1.11 with an unchanged shrouds diameter. Curve C is not fully acceptable but the "droopy" portion is replaced by a flat curve.

3.3 Third series of tests

Figure 3c shows prototype test results obtained with two different diffuser geometries and with an impeller similar to that of Fig. 3a curve C.

Curve A corresponds to a diffuser geometry not in accordance with the model design because of manufacturing tolerances. The diffuser channels inlet area was found to be smaller than the design value ; vanes inlet angle was also slightly uncorrect. These deviations produced a best efficiency point adapted to lower flows and a head flow curve with a steeper slope. The result was an unstable head-flow curve.

Curve B shows the head-flow curve obtained after correction of the diffuser geometry according to the model design.

This curve appears to be very similar to the one obtained in model tests (Fig. 3a curve C). The corresponding impeller-diffuser combination has been retained in tests presented below.

4 - DYNAMIC BEHAVIOUR OF THE PUMP AT DESIGN AND OFF-DESIGN CONDITIONS

4.1 Test facility and instrumentation

The pump has been instrumented in order to determine the impulses originating from it, particularly at low frequency. The identification of these excitations enables to design the pump baseplate as well as the junction with the piping system. The instrumentation comprised :

- 4 piezo electric strain-gauges, to measure the vertical and horizontal forces as well as the torque created by the pump on its pedestal.

- vibration measurements, close to the suction flange and discharge flange of the pump.

- two sets of three piezo electric pressure transducers, which have been mounted nearby the suction and discharge of the pump together with two hydrophones, in order to measure the pressure fluctuations. They enable to separate the fluctuations originating in the pump from the fluctuations coming from the circuit.

- One pressure transducer, close to the impeller outlet, to measure the pressure fluctuations within the space between the casing and the shroud of the impeller.

A balance-motor allowed to calibrate the strain gauges and to characterize the pump/circuit junction in the frequency range 3-20Hz.

4.2 Tests results

The test program was initated on the 0.73 scale model and pursued on the full scale prototype. The dynamic behaviour was evaluated over the whole flow range of the pump, from very low flow rates (33 %) - and even at minimum flow (0 %) - up to maximum capacity (134 %), at a rotational speed of 1500 RPM. An analysis of the results by frequency bands will be presented first, followed by a study of the characteristic spectrum lines.

4.2.1 - Analysis by frequency bands

The variation of the horizontal (F_H) and vertical (F_V) forces and of the torque (CO) is plotted in Fig. 4 as a function of the flow. From these results it appears that the energy of the FH spectrum is almost fully contained in the 2-50 Hz frequency band, while the energy of the CO spectrum is concentrated in the 2-24 Hz band. The vertical force F_V spectrum increases continuously from the 2-24 Hz band to the 2-160 Hz band. The F_V spectrum involves characteristic lines over the whole frequency domain.

From the analysis of the pressure fluctuations it is observed that, for the most part, the energy of the spectrum is contained in the low frequency band (3-40 Hz).

It is worth noting that all these curves show the same evolution for all the frequency bands, namely a minimum at the design flow. Fig. 5 shows the pressure fluctuation at the suction and at the discharge of the pump, referred to the static head of the pump at the corresponding flow, in the 0-250 Hz frequency domain. It may be seen in Fig. 5 that the pressure fluctuations are below 1 % of the static head within the normal flow range.

4.2.2 - Analysis of the frequency lines

The spectral analysis of the force fluctuations reveals :

- a mean level of excitation at low frequency

- two characteristic frequencies, the first one at the speed of rotation, the second one associated with the impeller blade passage frequency.

The spectrum of force fluctuations has several peaks at low frequency (4,5 Hz ; 9.2 Hz ; 14,5 Hz), which were already pointed out during the tests of pump excitation by the balance motor. However, as will be explained below, these excitations are not linked with the pump. In order to determine the influence of the circuit on the pump, a test at "zero flow" was performed. To this end the pump was disconnected from the piping system and coupled to the circuit by means of small flexible tubes. The comparative study indicates that the force fluctuations of the pump, mounted with the flexible tubes, are considerably lower than the ones measured with the normal piping assembly. Therefore it is concluded that these resonance frequencies are due to the circuit and not to the pump.

At the frequency of rotation (25 Hz) the force and torque fluctuations, corresponding to this line, are almost constant for all flow rates. This may result from a small unbalance of the impeller and is therefore independent of the operating conditions of the pumps.

At the impeller blade passage frequency (150 Hz) the energy contained in this spectrum line represents only 15 % of the total energy of the complete spectrum, as far as the horizontal force F_H and the torque CO are concerned. However, for the vertical force Fv, the energy level corresponding to this line reaches up to 50 % (at a flowrate of 330 l/s) of the total energy of this spectrum. Fig. 6a gives the evolution of the energy contained in this line versus flow-rate. It is seen that a huge variation takes place between 296 l/s and 330 l/s. This is due to the internal recirculation in the impeller eye, which has a detrimental effect on the vertical pulsations.

The hydro-acoustic power levels of the pressure fluctuations, corresponding to this line, are drawn in Fig. 6b, respectively at the suction and at the discharge of the pump. Recirculation appears clearly between 300 l/s and 330 l/s.

3 - PERFORMANCE PREDICTION AT 5000 RPM

Summarizing these findings, an overall frequency spectrum has been established : for each frequency the corresponding maximum value obtained during these tests has been determined.

On the basis of these frequency spectra of excitations originating from prototype tests at 1500 RPM, the dynamic characteristics of the pump under real working conditions (5000 RPM) have been predicted, by multiplying the measured values by $(5000/1500)^2$. This assumption has to be checked and will be qualified by laboratory tests as well as site tests.
The predicted overall frequency spectra of the pump operating at 5000 RPM are given in Fig. 7.

5 - CONCLUSIONS

The hydraulic performance tests, carried out with both the reduced scale model and the prototype, have confirmed the substantial improvement of the NPSH curves, corresponding to incipient cavitation, in part load operating conditions. It has also been shown how a stable head/flow curve may be obtained, despite a high value of the radial gap between impeller and diffuser.

Further, the effect of pumping operation at low flow-rates or at flow with internal recirculation on the forces transmitted by the pump to its pedestal has been substantiated.

In particular, force and pressure pulsations arising from impeller blade passing phenomena, have been highlighted, as well as low frequency pulsations originating from the circuit.

The experimental data allowed to establish a set of overall frequency spectra of excitations, originating form the pump and to predict its dynamic behaviour at nominal speed.

From these tests a fully reliable operating BFP has been developed, despite the high specific energy of this pump and the high constraint which resulted from part load pumping operation.

REFERENCES

(1) MAKAY E, NASS D, Gap-narrowing rings make booster pumps quiet at low flow - Power, September 1982

(2) MAKAYE, SZMODY O., Recommended Design Guidelines for Feedwater Pumps in large power generating units - EPRI Research project 1266-18, September 1980.

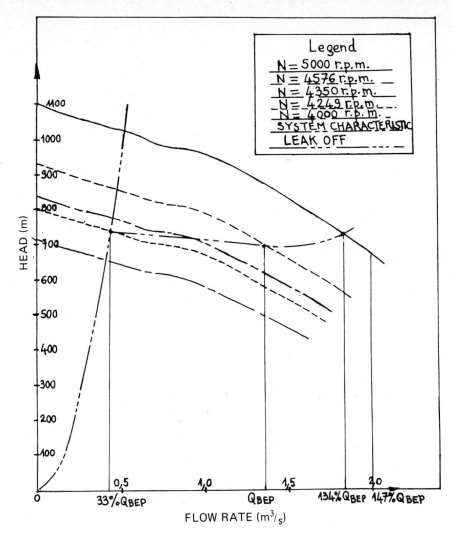

Fig 1   N4 main BFP working characteristics

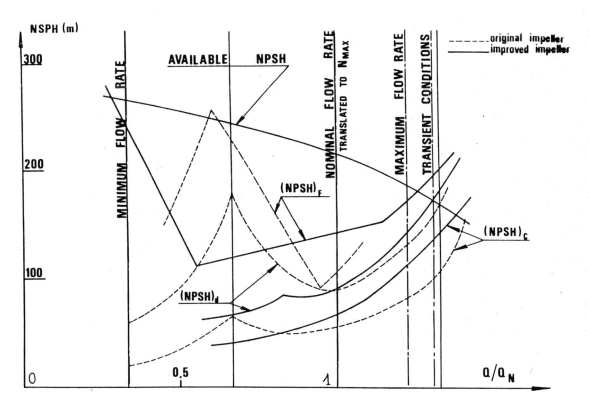

Fig 2   Cavitation performance

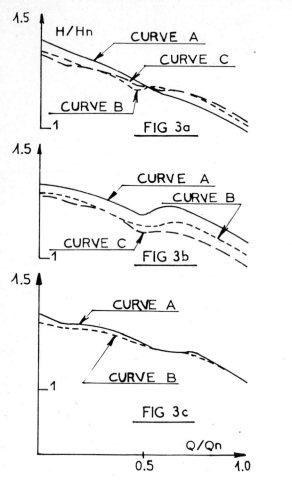

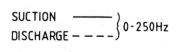

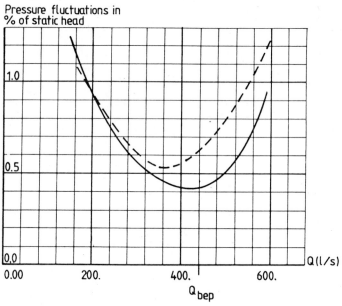

Fig 3   Head—flow curve stability

Fig 5   Analysis of pressure fluctuations by frequency bands

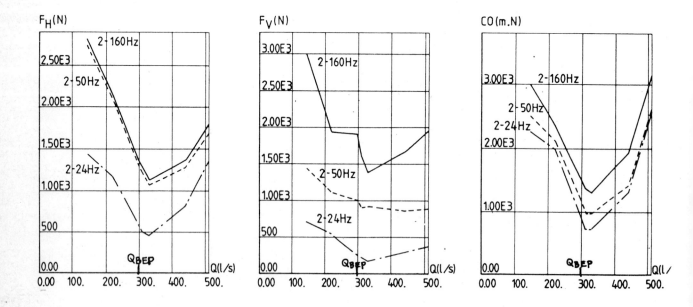

Fig 4   Analysis of forces fluctuations by frequency bands

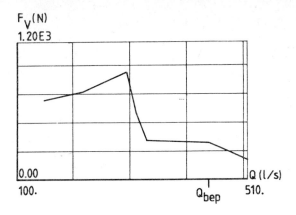

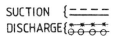

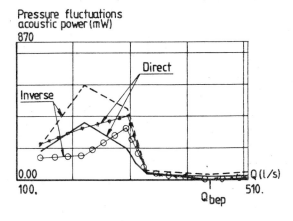

Fig 6   Analysis of forces and pressure fluctuations at
         blade passing frequency

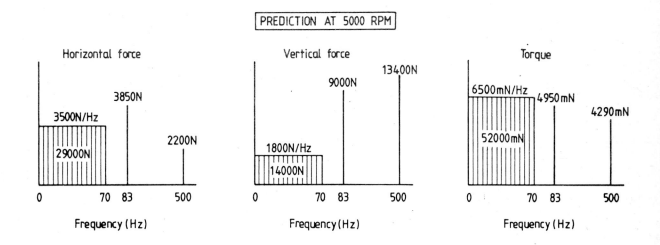

Fig 7   Forces overall spectra prediction at actual rotational speed

C345/88

Unstable behaviour of a centrifugal pump operating at part-load

B C STARK, BSc and **G TAYLOR**, BEng, PhD, CEng, MIMechE, MIOA
Imperial Chemical Industries plc, Billingham, Cleveland

SYNOPSIS This paper describes practical experience of unstable operation of a centrifugal pump in which the delivery line pressure fluctuated by +/- 60 bar about the mean line pressure of 180 barg. The reasons for the unstable behaviour are discussed together with the method used for analysing the system response and the pump modifications carried out to achieve stable operation.

1 INTRODUCTION

Unstable behaviour of centrifugal pumps operating at part load is not a new phenomenon but many users and vendors are unaware of the dangerous severity of pulsations and vibrations that can arise through unstable operation. The experiences detailed below relate to a particularly severe case in which the pump could not be operated safely. In less extreme cases, pump instability may not be recognised as the source of control and instrumentation problems and also of excessive maintenance.

2 PUMP SYSTEM

The pump is a 10 stage centrifugal pump designed to deliver 150 m^3/hr of water at a head of 1730 metres. A pump characteristic curve with a continuously falling head for increasing flowrate (negative gradient) was originally specified together with a maximum head of 1930 metres in order not to exceed the pressure limitations of the pipework system. It was recognised that this specification might pose hydraulic design problems but the performance requirements were not dissimilar to existing pumps in the system. The characteristic curve for the pump (as supplied) is shown in Figure 1. Note that the characteristic curves are scaled from tests carried out at approximately 2950 rpm to the design operating speed of 3550 rpm. The unacceptable shape was not recognised by the inspectors.

The pump operates in parallel with several other pumps. Water from a common supply ring main is discharged to a common delivery ring main. The total system flow is determined by individual take offs from the delivery ring main and thus the flow through the pump can vary widely although a minimum flow of 44 m^3/hr is maintained by means of a 'kickback' orifice.

The system is illustrated schematically in Figure 2.

3 OPERATING EXPERIENCE

For the commissioning trials the pump delivery was blanked off after the non-return valve, point B in Figure 2. When the pump was started both the delivery and suction pipework vibrated violently and the pump had to be shut down immediately.

The positive gradient on the characteristic curve in the 0 to 90 m^3/hr region was recognised as a source of instability which could create a resonance between points A and B in Figure 2 i.e. between the connection to the supply manifold and the closed end of the delivery line. To confirm this hypothesis dynamic pressure transducers were fitted to the suction and delivery lines of the pump and the pressure signals recorded on both an oscillograph and a tape recorder. The oscillograph traces displayed a regular sinusoidal pressure/time relation with a frequency of 11 Hz. The suction line frequency was also 11 Hz but there were additional harmonic components as the pressure fluctuation in the suction line was not purely sinusoidal due to the effects of cavitation.

Before embarking on pump modifications further trials were carried out with a longer delivery pipe. The internals of the non-return valves were removed and the delivery pipe blanked off approximately 50 metres from the pump, B' in Figure 2. Again the pump had to be shut down almost immediately after start up due to dramatic vibration not only of the pipework but also of the plant structure. The delivery pressure/time trace is shown in Figure 3 where it can be seen that the peak to peak pressure fluctuation in the delivery line was 120 bar at a frequency of 4.6 Hz.

The change in frequency due to the longer delivery line confirmed that the excessive pipe vibration was not due to a low energy level excitation of the mechanical natural frequency of the pipework and confirmed pump instability as the source of the vibration.

When the kickback orifice was increased to 100 m^3/hr capacity the pump operated in a stable manner.

4 PUMP OPERATION WITH A POSITIVE GRADIENT CHARACTERISTIC

The progress of a pressure wave through a simple pipe system (reservoir-pipe-valve) due to sudden valve closure is well known and documented in almost every text on 'water hammer'. The reflections of the pressure wave both at the open end of the reservoir and at the closed end of the valve lead to cyclic pressure fluctuation.

A pump inserted between the reservoir and the valve can be considered, at its simplest, as a device which, for a given flow, creates a pressure difference between the suction and the delivery line. The surge wave resulting from a valve closure passes through the pump. However for normal pump operation (i.e. operation in a region where the gradient of the pump characteristic is negative) then the reduced flow resulting from the surge leads to a higher pump differential pressure which offsets the pressure increase due to the surge. Thus the magnitude of a surge pressure wave is diminished when it passes through a pump operating with a normal characteristic curve.

The reverse is true when the gradient of the pump characteristic is positive. In this case the water hammer pressure wave is amplified. If the same reasoning is applied to the wave reflected from the reservoir then it can be seen that this also is amplified on passing through the pump and so on. Thus operation on the positive slope region of a pump characteristic is unstable.

The frequency of the pressure fluctuation is determined by the time for the waves to travel between the points of reflection but note that two traversals need to be completed for a full cycle. The frequency is thus given by the wave velocity divided by 4 * the length of pipe i.e.:-

frequency = c/4L

where:-
  c = wave velocity
  L = pipe length

In the present case the main wave reflections occur at the connection to the suction manifold (open end) and the closed end of the pipe. In acoustic terms the frequency corresponds to the resonant frequency of a quarter wavelength tube.

The wave velocity c is the speed of sound in the fluid in the pipe. It is slightly lower than the speed of sound in an infinite medium due to the flexibility of the pipe wall. Assuming the wave velocity to be 1200 m/s then the frequencies given by the above relation are 11.1 and 4.3 Hz which are close to the measured frequencies.

5 THEORETICAL SIMULATION

The unstable operating conditions can be demonstrated using any surge analysis program which has the capability of modelling the real pump characteristic. The simulations reported here were carried out using the HYPSMOP programs developed by the BHRA Fluid Engineering Centre.

In the simulations a small flow disturbance was introduced in order to excite the instability. The pressure time trace for a point just upstream of the pump delivery flange is shown in Figure 4 where the instability is clearly seen to develop within fractions of a second. Note that the magnitude of the pressure oscillation is limited by the shape of the characteristic curve in this simulation. In practice non-linearities in the system such as suction line cavitation can also limit the magnitude of the pulsation.

6 STABILISING THE SYSTEM

If an orifice plate is close coupled to the pump then the combination may be treated as a single device and the effective pump characteristic curve becomes stable at lower flows. Figure 1 shows the resulting effective pump curves for a variety of different orifice sizes. Increased orifice loss results in a curve which is more stable at lower flowrates but also results in a larger loss of performance at higher flows.

Had it been possible to locate the flow control valve on the pump discharge then this would have resulted in a stable system.

Simulation with different orifice sizes confirmed that instability arose when the flow moved into the regions of the characteristic curve with a positive gradient.

7 PUMP MODIFICATIONS

Any system modifications which would stabilise operation such as fitting orifice plates or increasing minimum flow resulted in unacceptable efficiency losses. The pump manufacturer therefore embarked on a test and modification program to produce an acceptable performance characteristic. The pump had been designed with a special stiff shaft and hence there was no previous experience specific to this design.

Three significant effects on performance were established:-

a) The axial position of the rotor, as determined by spacers adjacent to the thrust bearing, had a significant effect on the performance curve. Accurate reassembly after maintenance would therefore be critical.

b) The head at low flows could be improved by trimming the diffuser blade tips.

c) Significant improvements could be made to the pump characteristic by reducing the radial clearance between the impeller and diffuser which was found to be unnecessarily high.

Modifications were carried out to gain the maximum benefit from the above effects. Rings were fitted to the internal diameter of the diffusers in order to minimise the gap between the impeller and diffuser. This improved the flow paths and reduced the sensitivity to rotor position.

A more stable characteristic was achieved by modifying the hydraulic design of the radial vaned diffusers but a loss of efficiency also resulted from this. However some compensation for the loss of efficiency was achieved by back filing the impeller blades.

The modifications resulted in the pump characteristic shown in Figure 5.

8 SITE TESTING

A series of carefully arranged site tests were carried out so as not to risk the operation of the site power station. The tests established that the pump would operate satisfactorily with flows as low as 53 m^3/hr. At flows around 50 m^3/hr however instability occurs although it takes several minutes operation before the effects increase to the point where the pipework begins to shake at unacceptable levels. It is not clear why operation is not possible down to lower flows but this may be due to inaccuracies in the tests, changes brought about by transfer of the pump from test bed to plant or due to inaccurate scaling to allow for the operation at 3550 rpm.

9 SHAFT MOVEMENT

An interesting feature of this type of pump is the use of a pressure balance drum and a thrust bearing (Figure 6). Thrust load varies with flow. It was intended that the pump should have axial balance at a flow less than the minimum hence resulting in a unidirectional thrust on the thrust bearing and no axial shaft movement. Axial probes were not fitted during Works tests and so axial movement was not detected. During commissioning, however, the shaft was observed to move from one thrust face to the other, a distance of 0.4 mm. Axial balance was found to be at 70 m^3/hr.

10 CONCLUSIONS

The head/flow characteristic of a centrifugal pump is stable where the curve gradient is negative. Many pumps however have a characteristic with a positive gradient at low flows. Operation at part load in these unstable regions can give rise to very dramatic dynamic effects which can damage both pump and pump system. In unmanned and normally instrumented systems the reasons for failure might never be known.

In particularly severe cases unstable operation can be dangerous.

When purchasing centrifugal pumps for duties which require part load operation or operation in parallel it is important to specify and test for a stable characteristic.

Dynamic modelling techniques can be used to predict the magnitude and frequency of instability and can be used to assess the value of corrective measures.

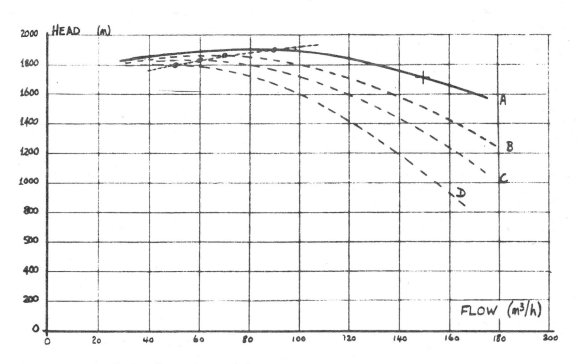

Fig 1   Pump characteristic curves
A   As supplied
B, C, D   Modified by fitting orifice to pump discharge
--●---●--   Minimum flow line for stable operation

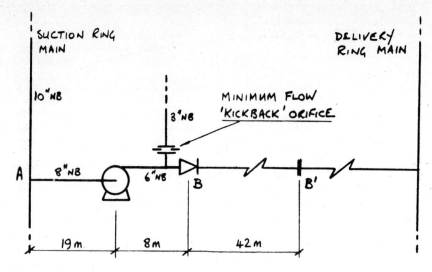

Fig 2   Schematic of pump system

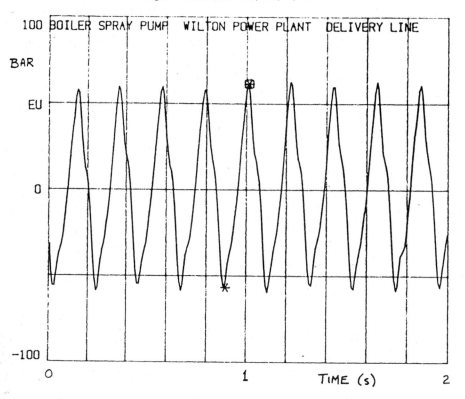

Fig 3   Dynamic pressure variation about mean pressure

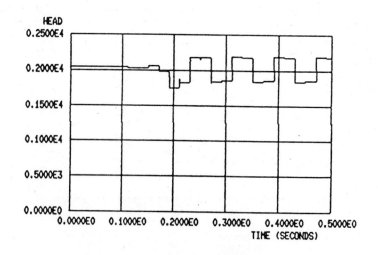

Fig 4   Simulation results; head versus time at pump discharge

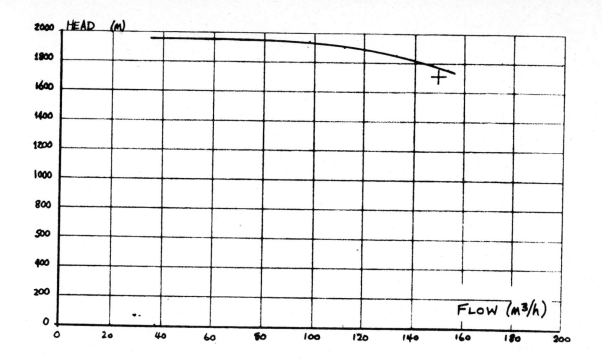

Fig 5   Pump characteristic after modification

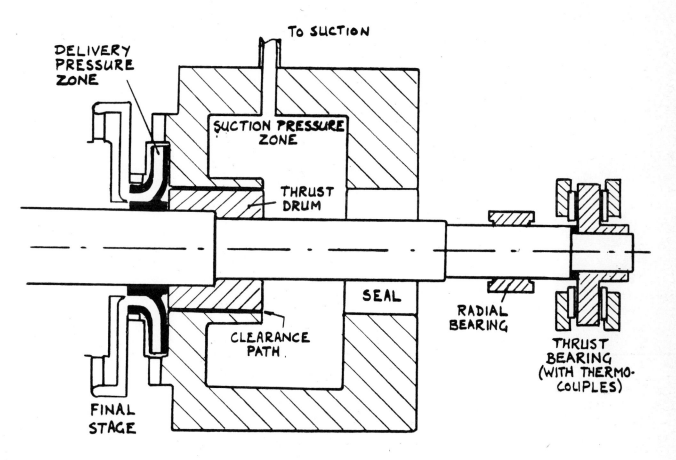

Fig 6   Pump balance drum and thrust bearing arrangement

C345/88 © IMechE 1988

135

C346/88

Measurement of the transfer matrix of a prototype multi-stage centrifugal pump

Y KAWATA, MSc, T TAKATA, MJSME, O YASUDA and T TAKEUCHI
Mitsubishi Heavy Industries Limited, Takasago City, Japan

SYNOPSIS A severe system instability occured at the partial operation of multistage centrifugal pumps. In order to prevent this low cycle instability at the next plant, a new design is adopted. The transfer matrix of this prototype multistage centrifugal pump is directly measured with the highly sophisticated method. The measured matrix up to 20Hz shows increase of unstableness as the flow rate decreased at the minimum flow rate. The unstableness is observed a little but this is negligibly small, and very small possibility of instability is expected. The pump test at the plant was carried out recently without any instabilities.

NOTATION

Aeq: equivalent area in the pump
f : frequency
Gij: transfer matrix
Hr: pressure ratio
Leq: pump equivalent length
p : pressure pulsation
q : flow pulsation
Z : hydraulic impedance
suffix
d : discharge
s : suction

1. INTRODUCTION

Severe system instability was experienced in the high pressure pipe systems with the multistage centrifugal pumps at partial operation. The instability was first recognized by the hammering of the discharge check valve. An example of measured pressure pulsation is shown in Fig. 1. This phenomena was not observed at the shop test and the pump Q–H curve shows clearly stable characteristic (negative gradient). The instabilities occurred only at partial flow rate with sufficient NPSH. The frequency was about 4.0Hz which was supposed to be the resonant frequency of the system.
The test using exactly the same 2 stage pump model showed that this pump has unstable dynamic behaviour at the certain frequency range of partial flow rate.[1][2]
This forced us to decide changing the design of the pump. The model tests using the same scale pump up to 3 stages were carried out to know the influence of the pump design parameter on the pump impedance. The result showed that the pump impedance becomes more stable when the design for the steeper slope of the stable Q–H curve is adopted. So this policy is adopted so for as the design requipment is fulfilled to this prototype pump. The dynamic test of the prototype pump was carried out to assure the effect of design improvement and to know the possibility of instability at the plant.

2. PUMP TRANSFER MATRIX

The pressure and flow pulsation at the pump suction and delivery is related by the transfer matrix in the frequency domain. The transfer matrix is composed of 2 x 2 complex numbers. This relation is shown by equation (1).

$$\begin{pmatrix} pd \\ qd \end{pmatrix} = \begin{pmatrix} G_{11} & G_{12} \\ G_{21} & G_{22} \end{pmatrix} \begin{pmatrix} ps \\ qs \end{pmatrix} \tag{1}$$

The G_{12} component is called pump impedance. The real part of G_{12} is called pump resistance and it has the same meaning as the slope of Q–H curve at 0Hz. This term is directly concerned with the unstable behaviour of the pump, as is easily related to the surging with positive slope of the Q–H curve. The main object of this research is to know how this term behaves when the pump operation condition is changed and how it is improved by the new design.
The imaginary part of the pump impedance is named pump inertance. This indicates the inertia effect of the water in the pump.
G_{11} and G_{22} are pressure and flow correlating terms respectively. They become 1.0 when the frequency approaches 0Hz. If cavitation exists in the pump, the term G_{22} differs from 1.0. G_{21} shows the pump compliance term which indicates the compressibility in the pump.

3. MEASURING TECHNIQUE OF TRANSFER MATRIX

3.1 Methodology of the measurement

The pump transfer matrix contains four complex unknowns. The test of one frequency at a certain boundary condition gives two equations using equation (1), so we need two sets of data if we want to obtain complete component of transfer matrix.
These are shown in equation (2).

$$\begin{pmatrix} pd & pd' \\ qd & qd' \end{pmatrix} = \begin{pmatrix} G_{22} & G_{12} \\ G_{21} & G_{22} \end{pmatrix} \begin{pmatrix} ps & ps' \\ qs & qs' \end{pmatrix} \tag{2}$$

where the dashed parameters mean the data tested with a different boundary condition. By the calculation we obtain each component of the matrix using equation (2).

$$G_{11} = \frac{1}{Zs-Zs'}\left(\frac{Zs}{Hr}-\frac{Zs'}{Hr'}\right)$$

$$G_{12} = \frac{ZsZs'}{Zs-Zs'}\left(\frac{1}{Hr'}-\frac{1}{Hr}\right)$$

$$G_{21} = \frac{1}{Zs-Zs'}\left(\frac{Zs}{HrZd}-\frac{Zs'}{Hr'Zd'}\right)$$

$$G_{22} = \frac{ZsZs'}{Zs-Zs'}\left(\frac{1}{Zd'Hr'}-\frac{1}{HrZd}\right) \qquad (3)$$

where Hr=ps/pd(pressure ratio)
    Z=p/q (impedance)

If we exchange suffix of equation (3) between s
and d, we obtain another set of equations. The
former is named suction based equation and the
latter delivery based equation. These two sets
of equations must give the same solution if the
measurement is sufficiently reliable. These
equations are solved with the computer and the
four components of the matrix are obtained.

3.2 Pump test loop for the measurement

The test loop is arranged by utilizing the shop
test loop. This loop is accomodated with the
booster pump and cooler at the outside of the
building.
The flow from the booster pump is stilled in the
suction tank and it goes through suction
measuring section, pump, delivery measuring
section, flow control valve and goes to the
cooler outside.
The oscillator with hydraulic system is attached
between the suction tank and suction measuring
section.
One of the greatest problems was whether this
high pressure system could be excited or not.
For this reason a very large actuator was
prepared. The capacity of the actuator is 2 x
10^5N with the hydraulic pressure 18MPa. By this
actuator we had enough excitation for the
measurement.
As the pump flange is directed upwards, the
pipes 6.5m long for the suction and 8.2m long
for the delivery are contained between the
measuring sections and the pump flanges
respectively.
As the pump is high head multistage pump and the
objective frequency is low, it is difficult to
establish two definitely different boundary
conditions. To achieve this, the switching
valves, the pipe line of 70m long and the
accumulator are attached to the loop.
Especially the pipe line is designed to furnish
2 valves at both ends and attached to the loop
so that this can act as a long main line or a
branch.
At the preliminary test, every combination is
examined by excitation and the difference of the
impedance is checked. As the result, 3
conditions are selected. From two sets of
measured data, one result is obtained using
equation (2). Therefore three results are
obtained out of three measured sets of data.
Out of these, two good results are selected.
This procedure is carried out with the suction
flow pulsation based case and delivery flow
pulsation based case. So four sets of data are
obtained by one measurement.

3.3 Measuring system

Fig. 3 shows the outline of the measuring
system. The pump and pipe line is excited from
2.0Hz to 20Hz with the increment of 0.5Hz, and
the data are stored in the computer. This
procedure is repeated three times with different
boundary conditions.
The main feature of this system is to introduce
vector filters and to form a closed loop control
system on the following components:
    fluid exciter → a pressure transducer →
    vector filter → computer → function
    generator → servo amplifier → fluid exciter
The signals of one pressure transducer are
compared with the set up value and the negative
feed back signals are sent to the function
generator to adjust the pressure pulsation
level. The test was carried out by adjusting
the suction pressure pulsations to avoid the
unsteady cavitation and to prevent the
unexpected damage caused by it.
Once the start and end frequency with a
frequency increment is set, the upper procedure
moves on automatically. This procedure is
repeated three times on the different boundary
conditions and after this the data are processed
by the computer.
This method works well with following
advantages; test time saving, the increase of
the data precision and the ability of measuring
with smaller frequency increment.[3]

4. TEST PUMP

The tests are carried out with a prototype ten
stage, opposed impeller type centrifugal pump.
The specifications of the prototype pump is
shown in Table 1.
This pump is heavy duty double case type and has
balance sleeve to reduce axial thrust. The
leakage flow from this clearance is returned to
the suction side through the balance pipe.
The rated flow rate is 159 m^3/h and the flow
rate at the maximum efficiency Q_{max} is 205
m^3/h. The minimum flow rate is 25 m^3/h which is
about 12.2% of the Q_{max}. The pump resistance
at this minimum flow rate is especially
important, because the phenomena at the plant
became severer at lower flow rate.

5. TEST RESULTS

The tests are carried out to know the influence
of flow rate and rotational speed. The results
of the tests are summerized in the following
sections.

5.1 Experimental transfer matrix

Typical measured pump transfer matrices are
shown in Fig. 4 as functions of frequency in
which the real and imaginary parts of each four
elements are shown. It shows the real part of
G_{12} becomes positive above 12Hz. But this
doesn't mean the pump is unstable above 12Hz,
because this measured transfer matrix includes
the influence of suction and delivery pipelines.
To get the correct pump transfer matrix, the
effect of these pipes must be eliminated.
The relation between the measured transfer
matrix and the correct pump transfer matrix is
shown in Fig. 5.
The results by this procedure are shown in Fig.
6.
Fig. 6 clearly shows the following
characteristics:
(1) This result is quite reasonable and the
    measurement was carried out very

accurately, because of the following reasons.
- The transfer matrices calculated from the different boundary conditions coincide very well as well as suction and delivery flow pulsation based cases.
- The value of real parts of G_{11} and G_{12} comes very near to 1.0 when the frequency approaches 0Hz. This agrees well as the theory.

(2) The imaginary part of G_{12} (inertance) exhibits a little depression about at 3Hz. Except for this depression it has almost the constant gradient at low frequencies. This gradient indicates the duct effect of the prototype pump. The relation between the equivalent pipe length Leq and cross sectional area Aeq is given as follows:

$$\frac{d(Imag(G_{12}))}{df} = \frac{-2\pi Leq}{g \cdot Aeq} \qquad (4)$$

From Fig. 6 the duct effect is calculated as Leq/Aeq=2300(1/m). This gives fairly large equivalent length compared with the physical pump and pipe dimensions.

(3) The real part of G_{12} (resistance) also shows a sharp depression about 3Hz. Except for this, it goes down towards the negative sides as the frequency increases. This shows that the pump becomes more stable as the frequency increases.

(4) The real part of G_{21} (compliance) indicates the elasticity of the casing, the compressibility of water or air volume in the pump. It is nearly zero, so that we conclude pump cavitation does not exist in the pump because of the high suction pressure.

5.2 Effect of pump operation condition

Fig. 7 shows the effect of flow rate on the resistance. It is compared with 4 different flow rates. It shows the following clear tendencies;
- The resistance between 2Hz and 15Hz changes greatly as the flow rate changes. This result obviously shows the greater unstableness as the flow rate decreased.
- The resistance between 15Hz and 20Hz is independent of the flow rate.
- At minimum flow rate 25m^3/h the small instability of pump is observed between 4.3Hz to 5.5Hz.
- The sharp depression at 3 to 4Hz is observed at every flow rate.

The following features indicate that this depression of the pump resistance is not the measuring error but the actual phenomena of the prototype pump.
- The result shows scarcely any scatter up to 20Hz.
- The result with different boundary condition shows exactly the same result.
- Only the G_{12} term (resistance and inertance) shows this tendency and the other element is not affected.
- This phenomena is commonly observed regardless of the pump conditions. This phenomena was not observed at the model test, with the same measuring

method[3], so it is assumed to be a unique one to the prototype pump.

Fig.8 shows the effect of rotational speed on the resistance. The tests are carried out by changing the rotational speed from 1800 r/min to 3600 r/min. In these tests the flow rates are adjusted to fulfil the same flow coefficient. The result clearly shows the following characteristics:
- The result also shows a steep depression about 3Hz as Fig. 7. This depression becomes steeper as the rotational speed is increased from 1800 r/min to 3000 r/min or 3600 r/min.
- The resistance shows a peak at the frequency just above the depression. Then it gets down at higher frequency region.
- These upswing and downswing are amplified as the rotational speed increases.
- These peak moves to higher frequencies as the rotational speed increases.

6. CONCLUSION

The transfer matrix of the prototype pump is measured accurately. By this test result we have the following conclusions.

(1) The pump resistance goes to unstable side as the flow rate is reduced. Also the resistance shows an unstable peak at low frequency. This explains why the system instability occurs only at the part load and its frequency is low.

(2) This tendency is emphasized as the pump rotational speed increases.

(3) The frequency range and the magnitude of the unstableness of the resistance on this newly designed pump is fairly small. This concludes the great improvement on the pump stability compared with the two stage pump[1], which had maximum resistance of 500 (s/m^2) at 2000 r/min.

(4) The calculated resonant frequency of the plant pipeline does not coincide with this frequency range. So that the instability is supposed not to occur at the plant.

The trial run of this pump was carried out at the plant recently and the system instability was not observed at all.

7. REFERENCE

(1) KAWATA, Y., EBARA, K., UEHARA, S., TAKATA, T.. The Dynamic Behavior of Centrifigal Pump and Its Effect on the System Instability. 2nd Meeting of IAHR W.G., Mexico City, Sep. 1985

(2) KAWATA, Y., EBARA, K., UEHARA, S., TAKATA, T.. System Instability Caused by the Dynamic Behaviour of a Centrifugal Pump at Partial Operation. JSME International Journal, 1987, Vol. 30, No. 260, 271-278

(3) KAWATA, Y., YASUDA, O., YOSHINO, H., UEHARA, S., TAKATA., TAKEUCHI, T.. Experimental Research on the Measurement of the Dynamic Behaviour of Multistage Centrifugal Pump. 3rd Meeting of IARH W.G., Lille, Sep. 1987.

Flow Rate	159m³/h
Pump Head	1000m
Rotating Speed	3000r/min
Specific Speed	19.9r/min·m³/S·m
Suction Diameter	$\phi 150 \times 10^{-3}$m
Delivery Diameter	$\phi 100 \times 10^{-3}$m

Table 1 Specification of the prototype pump

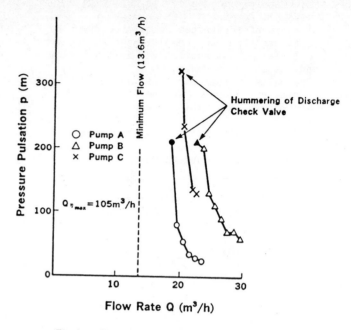

Fig 1 Example of pressure pulsation at a pump

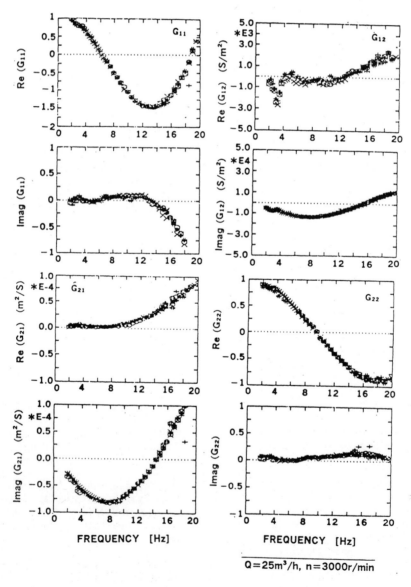

$Q = 25m³/h, n = 3000r/min$

Fig 4 Measured transfer matrix of prototype pump

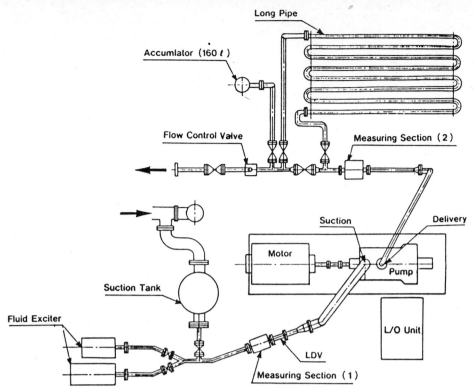

Fig 2   Test loop for measuring the dynamic behaviour of the
prototype multi-stage pump

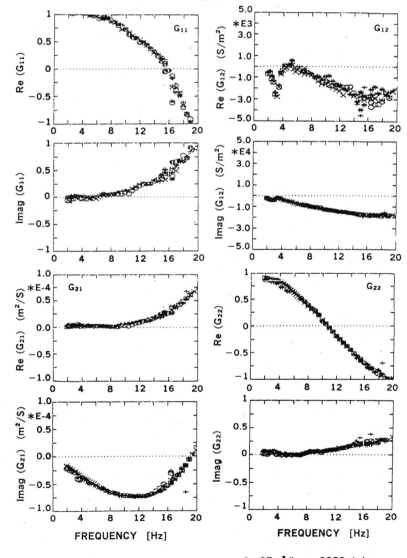

Q=25m³/h, n=3000r/min

Fig 6   Corrected transfer matrix of prototype pump

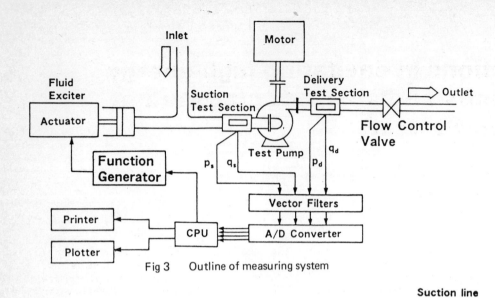

Fig 3   Outline of measuring system

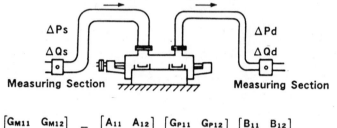

$$\begin{bmatrix} G_{M11} & G_{M12} \\ G_{M21} & G_{M22} \end{bmatrix} = \begin{bmatrix} A_{11} & A_{12} \\ A_{21} & A_{22} \end{bmatrix} \begin{bmatrix} G_{P11} & G_{P12} \\ G_{P21} & G_{P22} \end{bmatrix} \begin{bmatrix} B_{11} & B_{12} \\ B_{21} & B_{22} \end{bmatrix}$$

Measured	Suction line	Pump	Delivery line

$$G_M = A \cdot G_P \cdot B$$
$$G_P = A^{-1} \cdot G_M \cdot B^{-1}$$

A,B : Transfer Matrix of pipelne

Fig 5   Influence of suction and delivery pipe

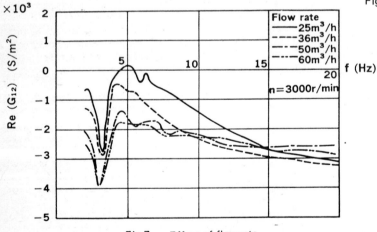

Fig 7   Effect of flowrate

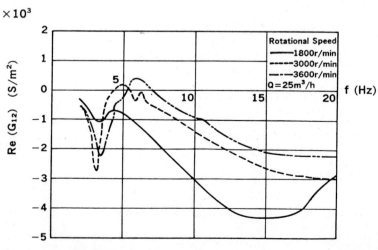

Fig 8   Effect of rotational speed

C347/88

Pressure pulsations in cavitating high-energy centrifugal pumps and adjacent pipework at very low or zero flowrate

E GRIST, BSc, PhD, CEng, FIMechE, FASME, MBIM
Central Electricity Generating Board, Knutsford, Cheshire

SYNOPSIS   A pressure pulsation phenomenon associated with the operation of cavitating high-energy centrifugal pumps at or near zero flowrate is described and an approximate analytical method presented which enables the possible consequences of its occurrence to be assessed. These are shown to be unusually high inlet pipe pressure and extremely rapid discharge non-return valve opening.

In practice very low or zero flowrate operation results when pump low-flow protection fails to operate correctly.   The phenomenon described relates therefore to a fault condition. A limited amount of evidence from power station operational experience is described in which such fault conditions have occurred.   Conclusions drawn indicate the prudence of taking the phenomenon into account when dealing with high-energy centrifugal pumps which might have to function at very low or zero flowrate.

NOTATION

H	rate of heat input to fluid from pump power
h	sensible heat after time Δt
K	bulk modulus of liquid in volume V_1
L	latent heat of vapourisation after time Δt
m	mass of fluid in volume V_2
NPSH	nett positive suction head (see Ref 5)
P	pressure
P_1	inlet pipe pressure (initial value)
P_2	discharge pipe pressure before non-return valve
P_3	discharge pipe pressure after non-return valve
dp	change in inlet pipe pressure
ΔP	generated pressure
P_{vap}	vapour pressure
Q	flowrate
q	dryness fraction
T	temperature
T_1	temperature of fluid (initial value)
t	time
Δt	increment in time
V_1	volume of pump inlet and discharge pipework ($V_1' + V_1''$)
V_2	swept volume of impeller blades and associated 'washed' volume ($V_1 \gg V_2$) i.e. fluid volume to which heat is added from impeller losses
V_g	specific volume of vapour
V_f	specific volume of liquid
dv	change in volume

Suffices (Used in Fig 4 and Appendix B)

a	values appropriate to an unconstrained system
b	values appropriate to a constrained system

1 INTRODUCTION

Pump power during normal operation is used to generate the required discharge pressure in a given range of forward flows which, for economic operation, are usually close to the best efficiency point. Losses in this process are dispersed mainly as heat to the liquid passing through the machine and are measurable as a temperature rise from inlet to discharge branch. This temperature rise is, until wear increases internal clearances and affects hydraulic performance, essentially independent of time.

When a cavitating high-energy* pump is operated in a manner which results in the forward flow falling to zero or very close to it conditions arise in which the heat generated is not carried away and the temperature of the fluid rapidly builds up. Under such transient operating conditions a pressure pulsation phenomenon is predicted to occur as a result of the explosive volumetric growth of the cavitating fluid within the pump impeller. It will be shown that in high-energy pumps the phenomenon can lead to unusually high inlet pipe pressures and/or extremely rapid opening of the discharge non-return valve.

* 'High-energy' is defined in Appendix A.

2 CENTRIFUGAL PUMP OPERATING REGIMES

As flow is reduced from that at best efficiency three distinct flow regimes are possible; stable, unstable and transient. The pressure pulsation phenomenon which is the subject of this paper is particular to the transient regime.

A general characterisation of the three possible flow regimes, which all manifest themselves in the externally measured performance of a centrifugal pump, is presented in Fig 1.

In the stable regime flow is essentially unidirectional. When cavitation occurs at a particular flowrate cavities progressively develop from the impeller blade inlet until they collapse further along the impeller passages. Visual observation of this process confirms the overall stability of such flows.

The inlet pressure of a pump operating in the stable regime is constant as shown in Fig 2 as Point 1.

In the unstable regime strong recirculatory motions occur within an impeller and in adjacent inlet pipework during part-load operation. When cavitation is present the pressure pulsations originating from such hydraulic instabilities can be severe. This is exacerbated if inlet pipework and impeller inlet geometry are conducive to the propagation of recirculating cavitating flows. The amplitude of pressure pulsations is determined by the characteristics of the cavitation volumetric growth/collapse process within the impeller (NPSH dependent) and the pre-rotation (pump inlet pipework geometry dependent). Examples of this regime and efforts to quantify it by analysis have been presented by Massey (1), Fraser (2) and Palgrave (3). An attempt to define the unstable working range by pump design type has been made by Makay and Adams (4).

The inlet pressure of a pump operating in the unstable regime fluctuates continuously in a periodic manner typically as shown in Fig 2 as Point 2.

In the transient regime little or no volumetric flow interchange occurs between fluid in or near the impellers and fluid in the surrounding pipework. Practically all the energy put into the fluid in the impeller goes to overcome its hydraulic losses and this results in the temperature of this fluid rising continuously. Where the volumetric expansion associated with this temperature rise is constrained the inlet pressure of high-energy pumps rapidly builds up. The periodic recirculations in or near impeller inlets at or near zero flow and the vapour bubble collapse associated with migration of heated fluid from impellers into the relatively cool liquid in the inlet pipework region inevitably gives this phenomenon the appearance of a pressure pulsation which increases in magnitude with time.

The inlet pressure of a pump operating in the transient regime builds up as shown in Fig 2 as Point 3.

3   THE PRESSURE PULSATION PHENOMENON

3.1 A zero flow inlet pipe pressure pulsation model

A model of the pressure pulsation phenomenon at zero flowrate is now described. This is complemented in Appendix B by a mathematical analysis of the pressure and volumetric growth characteristics of an impeller sufficient to broadly demonstrate by numerical example the

likely consequences of operating a pump under such conditions. For simplicity the influence of inlet recirculatory flows has been omitted.

Consider a pump in which the through flowrate has just become zero. In this condition the power supplied to overcome hydrodynamic losses heats the liquid contained within a volume which approximates to little more than the swept volume of the impeller blades. In high-energy pumps the magnitude of power input is such that the temperature of this liquid rapidly increases. Consequential expansion of the liquid occurs which, if no forward flowrate is possible, results in the inlet pipework pressure increasing against either a closed inlet non-return valve or the liquid inertia in the inlet pipework.

In practice high-energy input pumps operating at very low or zero flowrate are cavitating. In multistage pumps this may be limited to the first stage impeller. The NPSH available to prevent cavitation is usually determined by the high cost of providing more than the minimum necessary to protect pumps from the occurrence of unacceptable cavitation erosion damage whilst operating at or near best efficiency point. As shown by Grist (5) the NPSH required at part-load is substantially greater.

Heat addition to the fluid in a cavitating pump impeller leads to volumetric growth at the vapour/liquid interface which is much more rapid than that in the equivalent impeller filled with liquid. The violent mixing of liquid and vapour which occurs in an impeller at zero flowrate ensures that heat transfer between liquid and vapour is not restricted by liquid thermal conductivity effects. This simplifying fact is however offset by (i) the restriction on vapour volumetric growth imposed by the increase in local vapour pressure associated with increases in local static pressure (ii) the reduction in power input as the vapour volume begins to reduce the pressure generation capabilities of the pump.

Consider the typical high-energy pump system shown in Fig 3 in which the discharge non-return valve and the leak-off recirculation protection valves are all closed. In a well designed system this would be a fault condition brought about by failure of the low-flow protection to operate. With no throughflow the inlet non-return valve is also closed. Heat input by the pump results in expansion of the 'trapped' volume of fluid ABC which in turn pressurises the liquid in the pump and inlet pipework so that it presses closed more firmly the inlet non-return valve.

Within the cavitating impeller the local vapour pressure has been reached by the liquid at the liquid/vapour interface. On Fig 4 this is Point 1. As further heat is added an increased volume of vapour is produced.

For many high-energy pump systems and for most vapours (and certainly for steam), long before dry saturated conditions are reached at Point 2, the additional volume produced has

© IMechE 1988 C347/88

exerted a self-pressurising effect on the system volume ABC. This results in Point 2' being reached at the new temperature T where the added volume dv causes a change in local vapour pressure.

The change in vapour pressure associated with the change in temperature from T_1 to T equals the change in static pressure dp. This appears, of course, as an increment of pressure in both the inlet pipework and the discharge pipework. In high-energy pumps this process continues until the static pressure change dp associated with increasing values of T raises the pump discharge pressure to the non-return valve lift pressure P_3.

The maximum value of dp occurs when the pump impeller becomes vapour locked. In this condition a reduced pressure rise is generated by the impeller and, for the pressure to rise to that of the discharge non-return valve, a pressure increment approaching the difference between inlet and discharge pipework pressures is produced. The inlet pipe pressure P_1 rises to nearly that of the discharge pipework pressure P_3.

At zero flowrate the vapour volume rapidly grows and with favourable system geometry could reach the point where the generated pressure begins to fall before pressure P_3 is reached. However, although pump input power may rapidly diminish as an impeller becomes vapour locked, even in this case a significant amount of heat continues to be added to the fluid in the impeller. At zero flowrate the change in heat loss in a centrifugal pump impeller is directly proportional to the change in density.

Using the model described above an inlet pipe pressure transient is forecastable. For single stage pumps the pulsation model applies directly. For multistage pumps a summation of the changes in each impeller cell, taking account of the progressive vapour locking of successive stages, is a necessary refinement.

3.2 Discharge non-return valve response

At the moment that the pressure transient reaches the value which lifts the discharge non-return valve (i.e. when P_2 reaches P_3) the rate of volumetric increase within the pump and inlet pipework has reached a significant finite value. As the non-return valve starts to lift in response to this volumetric increase the pressure build-up is relieved. The rate of relief is dependent upon the inertia and compressibility of the liquid in the discharge pipework. Where the contained volume beyond the discharge non-return valve is large, vapour volumetric growth in the pump and adjacent pipework continues against essentially pressure P_3. Inevitably however, whatever the discharge pipework system design, a pressure overshoot will occur. This pressure overshoot is directly proportional to the forces opening the discharge non-return valve. Where relief to the volumetric growth within the pump is insignificant extremely rapid opening of the non-return valve can ensue.

The instantaneous value of volumetric growth attained when non-return valve lift commences is very dependent on the characteristics of the operating system. For a variable speed feedwater pump system with a high static lift such as shown in Fig 5, failure of the leak-off protection system to prevent flowrate falling to zero leads to a number of possibilities. At pump speed (d) for example a pressure pulse dp in excess of magnitude C-D will lift the non-return valve. If conditions are such that the pump impeller becomes vapour locked before this occurs the pressure pulse can approach magnitude C-F.

3.3 Numerical examples

The general assumption on which the analytical model of the pressure pulsation phenomenon is based is that fluids expand when heated. However, the phenomenon only assumes importance when it leads to either (i) the associated pressure transient challenging the structural integrity of pressure containing plant items or (ii) the rate of relief of fluid flowrate transients being outwith the operating capability of the system components. To explore this further examples are presented using the equations and method given in Appendix B. For simplicity these examples assume no impeller inlet recirculation and no system leakage loss. They are based on a cavitating impeller pumping water in a system such as that described by Fig 3.

For most fluids the vapour/liquid specific volume ratio and the liquid bulk modulus reduce significantly with increasing temperature. The value of the initial liquid temperature T_1 could, therefore, be expected to have an influence on the rate of inlet pressure rise and on the volumetric growth rate attained when discharge non-return valve lift occurs.

(i) Data used in examples

Calculated values for inlet pipe pressure rise and instantaneous volumetric growth rate for a pump operating in a system initially containing water at 20 C, 150 C and 180 C have been derived for two values of power input, 2000 kw and 6000 kW. In the examples the swept volume of impeller blades and associated 'washed' volume V_2 is taken to be 0.01 m^3 and the volume of pump inlet and discharge pipework V_1 to be 10 m^3.

The static pressure and volumetric growth rate transients calculated using the approximate method given in Appendix B are presented in Fig 6 and Fig 7 respectively.

(ii) Discussion

The pressure rise curves shown in Fig 6 reflect the rise in vapour pressure associated with the temperature increase in the impeller fluid. This is principally effective in delaying rapid pressure growth from lower initial temperatures. Inevitably, whatever the initial conditions, as more heat is added the more the pressure rise continues towards the discharge non-return valve lift pressure.

Examining the curves for 2000 kW input power more closely, it is evident that with 20 C initial temperature, if leak-off recirculation commences within less than three seconds, the pressure transient will be limited to less than 10 bar. This time requirement is easily achievable if the leak-off valve is operable. As the initial temperature rises however the time for the leak-off to react is significantly reduced and for a 150 C initial temperature the same transient pressure change occurs in about one second. With 6000 kW input power at the same 150 C initial temperature this time is reduced still further to less than 0.4 seconds.

If the above pump were operating at zero flowrate such that Fig 5 pressure difference C-D equals 10 bar, failure of the leak-off system to open would result in a non-return valve opening 'blip' occurring on the timescales indicated. As Point D falls, either due to reduction in pump speed or vapour locking of the impeller, larger values of pressure rise take place in the inlet and discharge pipework.

Examining Fig 7 reveals how the phenomenon affects non-return valve operation. At a low initial water temperature of 20 C the instantaneous volumetric growth rate is very large, typically many times the pump design flowrate, so that if conditions arise in which the valve lifts rapid opening can occur if the downstream conditions are conducive to this. Under 150 C initial conditions the instantaneous volumetric growth rate has a much reduced value.

Reviewing the above it is evident that in the system described with its 1000:1 pipework/impeller volume ratio (V_1/V_2) zero flowrate operation with water initially at 20 C and 2000 kW input power poses a risk of rapid non-return valve opening if the pump is held just below discharge pipe pressure P_3 . However if a pressure only a few bars lower is reached the pump can withstand several seconds operation without this risk arising. In an initially hotter system (150 C or 180 C) the pressure rise due to 2000 kW input power is much more rapid. In these circumstances the lifting of the non-return valve is, in most systems, more likely and although the instantaneous volumetric growth rate is very much less than that at 20 C it is still comparable with the system design flowrate. Failure of the leak-off protection to respond promptly on 150 C / 180 C systems could result in rapid non-return valve opening. If a higher initial power is applied (e.g. 6000 kW as shown in Fig 6 and Fig 7) the rate of pressure rise increases approximately in direct proportion.

In the examples given the latent heat of vapourisation is large compared with the rate at which heat is added to the mass of fluid in the impeller and the vapour never becomes dry. Analysis of longer term operation at zero flowrate, not an objective of this paper, would have to take into account the thermodynamic processes associated with heating superheated vapour. It may be necessary to consider this aspect further when liquids other than water are pumped.

4 SYSTEM INFLUENCES

4.1 System leakage losses

System leakage losses will affect the rate of change of pressure and the time to reach non-return valve opening. From the volumetric growth rates given by numerical examples however it is clear that in high-energy pump systems this will often be a second order effect.

In practice, the main leakage loss from the contained volume ABC in Fig 3 is often that through shaft gland seals. In high-energy pumps with labyrinth glands which have injection systems a clear but small relief path can exist once the internal pump pressure exceeds that of the injected liquid supply. The resistance of flowrate offered by fixed bush labyrinths and their equivalents varies directly as the square of the leakage flowrate. In pumps fitted with mechanical shaft seals no relief path is provided until the seal fails. After failure, flowrate along the leakage path approximates to that of a narrow clearance fixed bush labyrinth.

4.2 Systems without inlet non-return valves

In inlet pipework systems of very low liquid inertia the pressure rise associated with operation at zero flowrate will be small since the volumetric expansion within the impeller can be rapidly accommodated by displacement of inlet pipe liquid. Typically this is the case in Works Test Facilities. In systems where inlet pipework is lengthy and the contained liquid has considerable inertia the pressure rise could be significant in terms of inlet piping design pressure limitations.

5 PUMP INFLUENCES

5.1 Impeller inlet recirculating flows

There is no reason why the periodic recirculation associated with unsteady part-load cavitating impellers should cease during transient operation. Given such an occurrence it is reasonable to assume that an additional volume of liquid is drawn into the recirculation within the impeller and becomes part of the heated volume V_2. The pressure generated by the pump in these circumstances is successively reduced as the vapour volume increases with each cycle through the impeller as shown in Fig 1.

5.2 Initial very low pump throughflow

Where a low flowrate is insufficient to carry heat away from the pump impeller as fast as it is produced through hydrodynamic losses transient conditions will arise. If, additionally, the volumetric growth rate of vapour within the impeller greatly exceeds this forward pump flowrate pressure build-up will again occur. This will only be relieved by the acceleration of liquid in the discharge line until either equilibrium is restored or the impeller vapour locks and the discharge pressure collapses closing the discharge non-return valve.

6   OPERATIONAL EXPERIENCE

It is never an intent to run high-energy pumps without low-flow protection and such pumps are therefore not normally instrumented to record such events. A single example of a large boiler feed pump at zero flowrate has however been recorded within CEGB during investigations into the causes of severe pipework vibration. As shown in Fig 8 the inlet pipe pressure on the two-stage feed pump grew rapidly in a pulsating manner until a pressure approximating to that of the discharge line (about 200 bar) was reached. This unexpectedly high pressure caused the 200 bar transducer used to suffer mechanical damage. The trace has the characteristics of a transducer being subjected to pressure pulsations beyond its range (hence the flats at 200 bar where it was electrically saturated) and can only be regarded as indicative at the high pressure end. Measures taken subsequently to protect the pump against zero flow operation prevented the test being repeated.

In PWR Power Station systems, many of which are similar in outline to Fig 3, several instances of mechanical failure in feed pump discharge non-return valves of the swing-check type have been recorded. The mechanism leading to the simultaneous failure of five swing-check valves at one Station (6) (7) has been described as being flow induced vibration which caused repeated impact between the disc stud and disc stop. In a second Station (8) where failure of the valve clamping assembly was attributed to corrosion, it could be significant to note that the leak-off recirculation valve to the fixed speed feedwater pump failed to open promptly as the pump came rapidly off load prior to this finding. At other Stations the design and/or assembly quality control of swing-check valves has been questioned following failures. It is suggested that at least in some of these cases the mechanism described in this paper might have played a part, perhaps a significant part, in the failures reported.

Whilst far from conclusive the limited experience of operation in the abnormal circumstances of zero flowrate gives some support for the need to pursue the consequences of the pressure pulsation phenomenon.

7.   CONCLUSIONS

1. A severe pressure pulsation phenomenon is shown to be possible in high-energy centrifugal pumps operating at very low or zero flowrate conditions.

2. It is prudent to evaluate the possible consequences of this phenomenon should failure of low-flow protection for such pumps occur. These are high inlet pipe pressure and/or rapid discharge non-return valve opening.

3. The behaviour of centrifugal pumps in the transient regime requires further and more detailed exploration.

4. A detailed analytical model which includes the effects of discharge system compressibility and inertia, impeller vapour locking and shaft seal losses could usefully be developed.

8   ACKNOWLEDGMENT

The Author wishes to thank the Central Electricity Generating Board for permission to publish this paper.

REFERENCES

(1) MASSEY, I. C. 'The Suction Instability Problem in Rotodynamic Pumps' Paper 4-1, NEL Conf. Pumps & Turbines Sept 1976.

(2) FRASER, W. H. 'Recirculation in Centrifugal Pumps' ASME Winter Annual Meeting pp 65-86 Nov 1981.

(3) PALGRAVE, R. 'Operation of Centrifugal Pumps at Partial Capacity' BHRA 9th Int. Tech. Conf. pp 57-70 April 1985.

(4) MAKAY, E. & ADAMS, M. L. 'Operation and Design Evaluation of Main Coolant Pumps for PWR & BWR Service' EPRI NP-1194 September 1979.

(5) GRIST, E. 'Nett Positive Suction Head Requirements for Avoidance of Unacceptable Cavitation Erosion in Centrifugal Pumps' IMechE Cavitation Conf. pp 153-162 1974.

(6) 'Loss of Power and Water Hammer Event at San Onofre Unit 1 on Nov. 21 1985' US Nuclear Regulatory Commission Nureg 1190 Jan 1986.

(7) CHIN, C. & REILLY, J. T. 'Plant Availability Improvement by Elimination of Water Hammer' ASME Paper 86-JPGC-NE-7 Oct 1986.

(8) US Nuclear Regulatory Commission Reports 50-280/86-42 and 50-281/86-42 Feb 1987.

APPENDIX A

DEFINITION OF HIGH-ENERGY

The description 'high-energy' is a relative one. In this paper it is shown that the effect of large rates of heat input into a pumped system can be significantly changed by system geometry and pumped fluid properties. For general considerations however 'high-energy' is taken to be a power input exceeding 100 kW per litre of impeller blade swept volume.

APPENDIX B

APPROXIMATE METHOD FOR EVALUATING
PUMP TRANSIENT PERFORMANCE CHARACTERISTICS
OF CAVITATING IMPELLERS

At Fig 4 Point 1 vapour is forming at the liquid boundary as heat is added. The following method may be used to obtain an approximate value of the inlet pipe pressure rise as the constraining effect of the system causes the vapour pressure to rise to that associated with temperature T.

The heat added by the impeller in time Δt to liquid at temperature T_1 would in an unconstrained system produce a small additional volume of wet steam from which the dryness fraction can be calculated.

heat added per unit mass of fluid in incremental time Δt.

$$= \frac{H.\Delta t}{m} = q_a \, L_a$$

Making the crude assumption that the mass of fluid in the heated zone remains constant (a good approximation in the first steps and not too erroneous at high values of T where the value of V_g approaches that of V_f)

Mass of fluid in heated zone

$$m = \frac{V_2}{V_{fa}}$$

Therefore

$$q_a = \frac{V_{fa} \cdot H \cdot \Delta t}{V_2 \cdot L_a} \qquad (1)$$

From Fig 4 assuming that the heat put into the remaining liquid can be neglected then:-

$$h_a + q_a \cdot L_a = h_b + q_b \cdot L_b$$

Giving

$$q_b = \frac{h_a + q_a \cdot L_a - h_b}{L_b}$$

Change in vapour volume

$$dv = \frac{V_2}{V_{fa}} \cdot q_b \, V_{gb}$$

Assuming again that only the liquid in the unheated zone V_1 (much larger than V_2) must absorb this increase in volume then:-

$$dp = \frac{K}{V_1} \, dv$$

Combining the above equations gives:-

$$dp = K \cdot \frac{V_2}{V_1} \cdot \frac{V_{gb}}{V_{fa}} \left(\frac{h_a + q_a \, L_a - h_b}{L_b} \right) \qquad (2)$$

Since liquid continues to evaporate to produce cavities at the new temperature T corresponding to the vapour pressure $P_{vap\,b}$ then:-

$$dp = P_{vap\,b} - P_{vap\,a} \qquad (3)$$

For thermodynamic equilibrium the values of dp in Equations (2) and (3) must be identical. The nature of fluid properties makes it necessary to solve these simultaneous equations graphically for each incremental step in time Δt to give values T, P_{vap} appropriate to Fig 4 Point 2'.

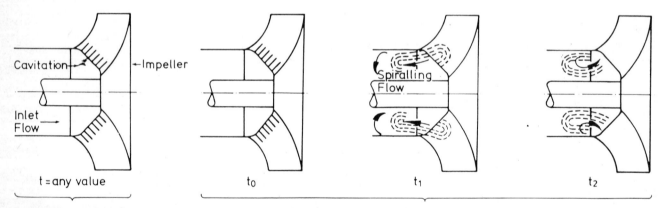

1. Stable Operation   2. Unstable Operation

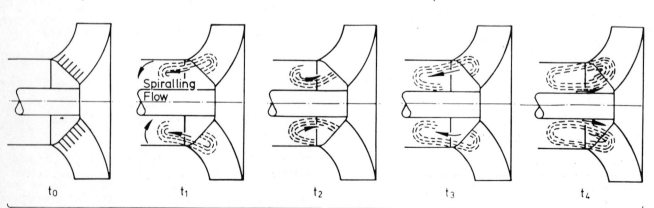

3. Transient Operation

Fig 1   General characterization of steady, unsteady and transient operation

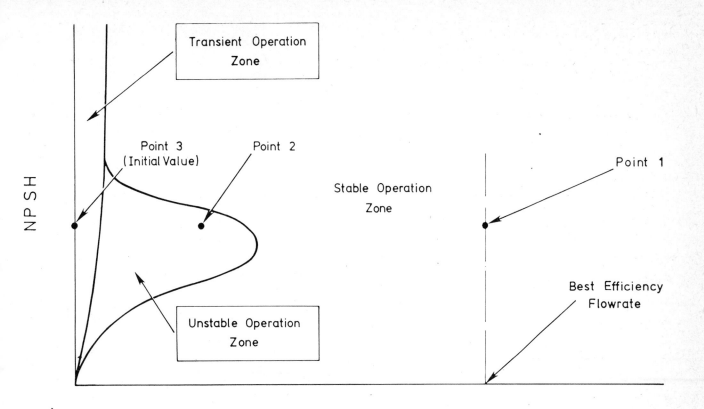

Flowrate Q

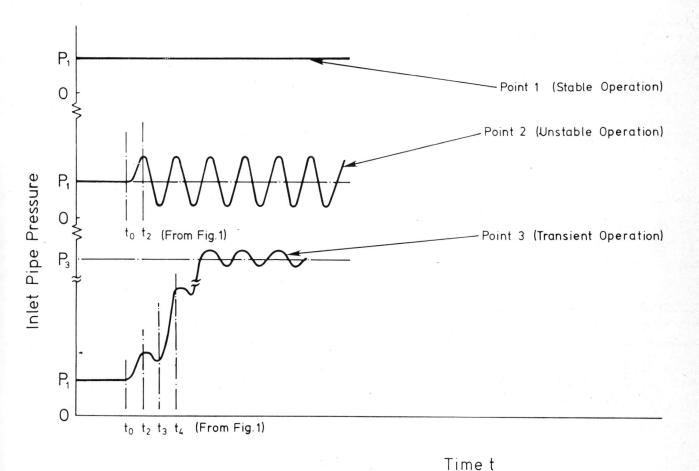

Time t

Fig 2   NPSH/flowrate and inlet pressure/time for steady, unsteady and
transient operation

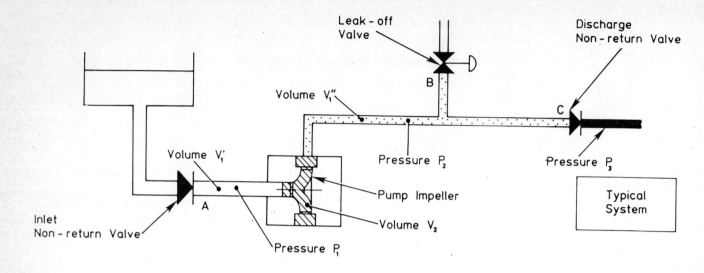

Leak - off
Valve

Discharge
Non - return Valve

Volume V_1''

B

Pressure P_2

C

Pressure P_3

Volume V_1'

Pump Impeller

A

Volume V_2

Inlet
Non - return Valve

Pressure P_1

Typical
System

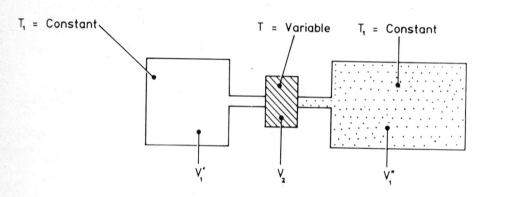

T_1 = Constant

T = Variable

T_1 = Constant

Equivalent
Volumetric
System

V_1'

V_2

V_1''

Fig 3   High-energy pump system layout (typical)

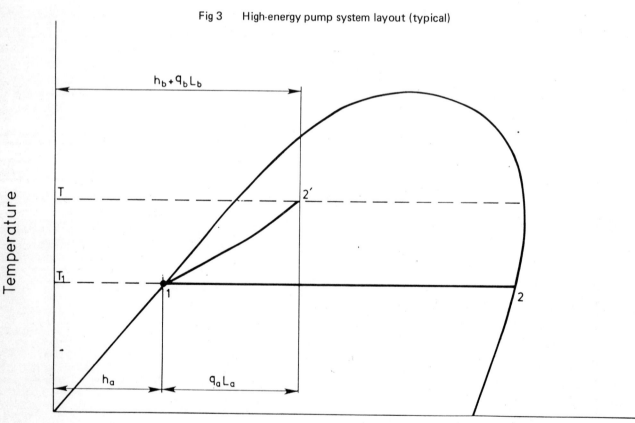

Temperature

Enthalpy

$h_b + q_b L_b$

T

2'

T_1

1

2

h_a

$q_a L_a$

Fig 4   Temperature/enthalpy diagram

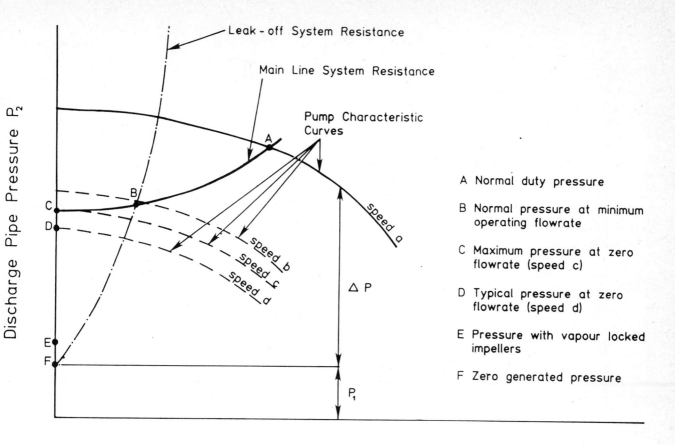

A Normal duty pressure

B Normal pressure at minimum operating flowrate

C Maximum pressure at zero flowrate (speed c)

D Typical pressure at zero flowrate (speed d)

E Pressure with vapour locked impellers

F Zero generated pressure

Fig 5   Pump/system characteristic diagram (typical)

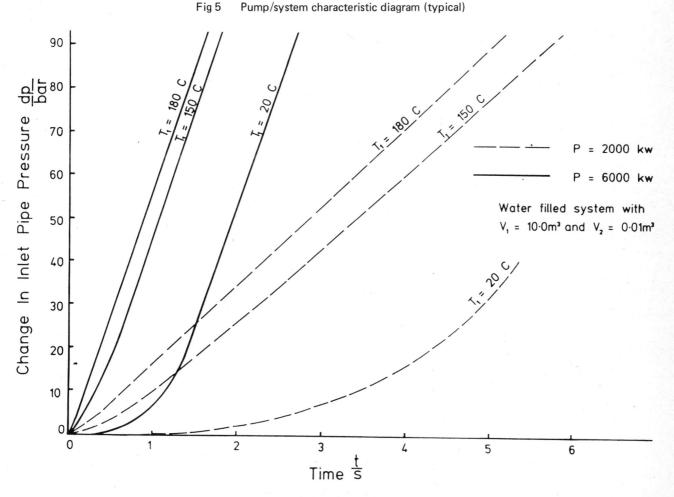

Fig 6   Inlet pipe pressure rise with time (calculated)

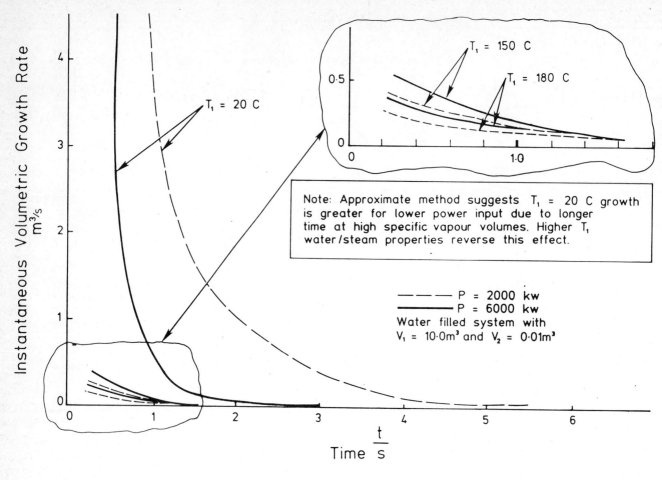

Fig 7   Instantaneous volumetric growth rate with time (calculated)

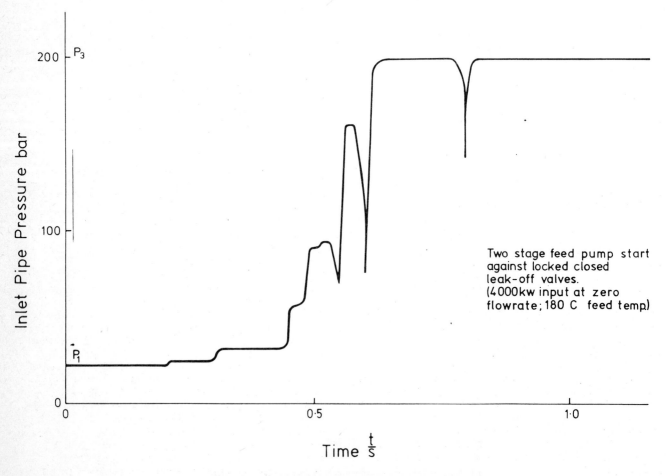

Fig 8   Inlet pipe pressure rise with time (recorded)

C348/88

Unsteady hydraulic forces in centrifugal pumps

J J VERHOEVEN
Byron Jackson BWIP, Etten-Leur, The Netherlands

SYNOPSIS. The forces that act upon a pump rotor stem from mechanical and hydraulic causes. The mechanical forces due to unbalance, misalignment, etc., are independent of flow rate and well known and limited by international standards and vendor/user specifications.

Unsteady hydraulic impeller forces are rather unknown and vary with flow rates. Since pump reliability is heavily related with vibration levels of the pump it is important to know the character and behaviour of hydraulic forces at the impellers generating these pump vibrations.
It is at present quite clear that vibration response of centrifugal pumps is mainly generated (excited) by the impeller hydraulic excitation forces instead of the mechanical forces.

This paper describes recent research results on unsteady hydraulic impeller forces. This data allows to define general ranges of the impeller hydraulic excitation forces, which is valuable data to judge pump operation and vibrations at BEP and part flow duties.
Practical ways to reduce hydraulic excitation forces are shown at the end of the paper.

NOTATION

Single variables

b_2	= impeller width at tip.
D_2	= impeller outside diameter.
$F_{(\omega)}$	= excitation force at frequency (ω).
F_r	= radial force.
$H_{(\omega)}$	= general transfer function for frequency component ω.
K_r	= non-dimensional impeller radial force factor.
m	= matrix size.
n	= matrix size.
Δp	= head rise generated by one impeller stage.
$X_{(\omega)}$	= response amplitude at frequency ω.
ω	= frequency.

Vectors

$\{F_{(\omega)}\}$	= fourier transformed excitation force vector.
$\{X_{(\omega)}\}$	= fourier transformed response vector.

Matrices

$[H_{(\omega)}]$	= transfer function matrix for frequency component ω.
$[H_{(\omega)}]^{-1}$	= inverse transfer function matrix for frequency component ω.

1 INTRODUCTION

Pump life time and reliability are heavily related to pump vibration levels, which are generated by mechanical and hydraulic excitation forces.
Unlike the mechanical forces, there is still a lack of knowledge on hydraulic excitation forces especially at run-out and part load flow rates (ref. 1, 2, 3, 4).

Hydraulic forces arise at the impeller in two distinctly different ways. Firstly when the shaft center is fixed, radial forces occur as a result of the distribution of static pressure and fluid momentum, around the impellers. These forces are termed excitation forces and are independent of rotor motion. Secondly, when the shaft center also moves (as a result of vibration) both at the impeller and the annular clearance seals additional forces are generated. These forces which are motion dependent can be represented by force coefficients in terms of stiffness, damping and mass. The latter forces received extensively research during the last decade (ref. 2) and are better known as "lomakin effect" in the annular clearance seals and impeller/casing "hydrodynamic interaction" forces. This resulted in quite effective dynamic models of the pump rotors, which can be used to simulate vibration response with special purpose computer programs. In practice even simple unbalance response analysis, using the maximum mechanical unbalance allowed according to factory standards, will never produce the results obtained during performance testing and field operation. It is at present quite clear that the vibration response of centrifugal pumps is mainly generated by impeller hydraulic excitation forces instead of mechanical excitation forces like unbalance, misalignment, disc skew, shaft bow, etc. A recently developed indirect method (5) allows prediction of impeller hydraulic excitation forces in short time with minimal costs. By using test stand vibration data and an indirect modal force prediction algorithm (5) 47 barrel type, multi stage pumps were analysed during the last two years. Also special tests are made on a large single stage process pump, to obtain data on the influence of hydraulic geometry and hydraulic deviations. This paper presents some of the results obtained, especially at part load and run-out conditions. Finally the paper summarizes ways to reduce hydraulic impeller forces at part load and rated flow conditions.

2 METHOD FOR DETERMINATION OF UNSTEADY HYDRAULIC IMPELLER FORCES

Instead of using test rigs with direct force measurement devices like strain gauges, load cell, etc., an indirect modal method is used to determine the hydraulic forces (ref. 5).

Analogous to experimental modal analysis, the indirect method uses frequency response functions which are determined between the locations of force excitation and the actual response measurement locations during performance testing.
The frequency response function is the linear relation between the response to the force input as a function of frequency.

In general : $H(\omega) = \dfrac{X(\omega)}{F(\omega)}$

If one considers a rotating machine as a complex mechanical system having m locations of force excitation and n locations at which response measurements are made, we can define a general linear matrix equation relating all excitation forces with all response measurement locations:

$$\{X(\omega)\}_n = [H(\omega)]_{nxm} \cdot \{F(\omega)\}_m$$

This matrix contains frequency response function, also called transfer functions, and is characteristic for the rotating machine itself and independent of actual excitation as long as the system remains linear. Once the transfer function matrix has been determined, the actual operating excitation forces can be calculated after inversion of this matrix. Multiplying both parts of the above matrix by the inverse of the transfer function matrix yields:

$$\{F(\omega)\}_m = [H(\omega)]_{mxn}^{-1} \cdot \{X(\omega)\}_n$$

By using actual operating vibration data for the response vector, measured simultaneously at all response locations for an excitation frequency (ω) of interest, the actual operating excitation forces are calculated. Normally, experimental modal analysis is used to determine the transfer functions by means of artificial excitation. Desanghere (6) and Okubo (7) developed excitation force prediction techniques using artificial excitation of complex mechanical structures. Artificial excitation of pump /turbine rotors over a wide frequency range, at exactly those locations where the real operating hydraulic excitation are working is very difficult, if not impossible. By using numerical excitation on a mathematical rotor model, allows to determine transfer functions between excitation locations and response measurement locations.
Figure 1 shows an overview of the combined analytical experimental prediction of hydraulic impeller forces using numerical excitation on a computer model of the rotor/stator and oper-

ating vibration data.
The proposed indirect modal method has certain limitations:

- Since the super position principle is used, the method will only work for linear rotor/stator systems.
- The accuracy of the mathematical model determines the accuracy of the predicted force levels. This means that modelling must be very accurate.

Advantages of the method are:

- Method allows generation of transfer functions over wide frequency range at any operating speed of the machine.
- No mounting problems or hydraulic disturbances arise by exciting typical hydraulic force locations.
- Excitation method can easily be changed numerically.
- Low research costs, since no special test rig is required.
- By selecting, numerically, correct phase angles for translatory harmonic force excitation either forward or backward whirling of the rotor can be selected. (This is experimentally very difficult.)

Bearing in mind the linear limitations and accurate modelling requirements as well as the possible error sources, this method allows to produce valuable hydraulic impeller force data in short research time. Experimental verification and detailed description of error sources, numerical inversion of transfer matrix, and numerical excitation are given in reference 5.

3 IMPELLER FORCE PREDICTION RESEARCH PROGRAM

The research at the authors company was focussed on large double case, multi stage, double volute pumps used in boiler feed, water injection and hydro carbon feed service.
Till present 47 pumps were subject to the force prediction algorithm, using a specially written computer program FIPC. This program contains all the analysis per figure 1.
Operational vibration data was taken during shop performance testing and subject to fourier analysis, using a modern FFT analyser. This fourier transformed vibration data is fed to the force prediction program, which in turn produces fourier transformed force frequency spectrograms.
Prior to performance testing, the pump impellers and casings were subject to carefull dimensional measurements, to obtain accurate data on hydraulic geometry deviations. The pump rotors were also dynamically balanced on a dynamic balancing machine. Residual unbalances were carefully registrated after balancing. Since the volute casings of the double case pumps are axially split, there will be no rotor disassembly necessary after balancing of the rotor. Residual unbalances registrated on the dynamic balancing machine were representing also the actual mechanical unbalance during performance testing. Both the dimensional measurements and residual unbalance measurements prior to testing allow correlation of the pure hydraulic impeller forces with following parameters:

- Pump flow rate.
- Hydraulic impeller/casing parameters.
- Geometrical tolerances in impeller/casing.

To illustrate the range and amount of data generated by this test program, it is necessary to consider that the 47 pumps contained 327 impellers, and the sizes of the pumps tested varied over a wide range:

* Impeller diameter   : 205 – 420 mm.
* Impeller specific speeds : 1000 – 2000
                              (US gal/min, ft, rev/min)
                            : 19.4 – 38.8
                              (m^3/HR, mtr, rev/min)
* Flow rates at BEP   : 50 – 1600 m^3/hr.
* Power ratings   : 250 –17000 kw.
* Number of stages   : 4 – 12.
* Operating speed for
  collection of force data : 1500 – 8900 rpm.
* Casing design   : Symmetrical and non symmetrical double volute designs.

Furthermore, the single stage process pump first used for verification of the indirect modal force prediction algorithm is used to study the influence of specific geometric deviations and hydraulic parameters. The pump is described in detail in reference 5. Prior to testing the impeller and casing were carefully measured for hydraulic dimensions and residual unbalance. The pump impeller forces were than predicted, for this original pump without any changes to the hydraulic waterways. Then several changes were introduced to the hydraulic geometry by using special metal polyester. Running the pumps with these changes of metal polyester, allows to determine the hydraulic impeller forces, which can be compared with the original design.
Also hydraulic parameters were varied by either test or design modifications or by rework of hydraulic dimensions. In total 14 different hydraulic deviations and hydraulic parameters were varied.

The force prediction algorithm FIPC used to correlate the force data produces typical excitation force frequency spectrograms representing a simulated fourier transform of the excitation forces. Figure 2 shows a typical impeller excitation force spectrogram produced by FIPC for the single stage process pump, subject to the special test program. It is clear from the force spectrogram, that the hydraulic forces have a very complex character. Analogous to other researches the hydraulic impeller forces are best characterized by their main frequency components:

* Low frequency component (1/10 – 1/5 x speed).
* Synchronous frequency component.
* Harmonics of operating speed.
* Vane passing force component.
* Random portions of force spectrum.

By using almost always numerical translatory excitation, the forces are predicted as two independent orthogonal forces at each excitation location. This allows to determine whether the forces are having a circular force character like unbalance or a specific eliptical character.

This is interesting data, since it can indicate specific locations where large hydraulic forces are generated especially at part load conditions.

The verification of translatory numerical excitation to predict translatory force components is shown by unbalance perturbation testing of the single stage process pump in reference 5.

4 IMPELLER EXCITATION FORCES VERSUS FLOW RATE

Excitation forces generated by an impeller consists of forces due to secondary flow fields, flow separation and by non-uniform pressure and momenta around the impeller. Synchronous impeller excitation forces, sometimes called "hydraulic unbalance" are generated in general by hydraulic deviations in the impeller waterway and non-symmetric deviations in the pump casing hydraulic geometry. This is the reason why these forces show a distinct difference with sub-synchronous and super synchronous (vane-passing) forces over flow rate.
The latter forces depend on the secondary flow fields, flow separation at volute tongue and impeller blades, and the blade/casing non uniform pressure interaction. These forces are therefore more related to hydraulic parameters and hydraulic design. They vary over flow range, with large values at reduced flow and run-out flow conditions. Around pump BEP flows they tend to minimum values. The synchronous forces, which are more related to geometric deviations have a more constant character over flow range, especially at moderate operating speeds. The results of all 47 multi stage pumps tested in the shop, allows to define typical ranges for these hydraulic excitation forces. To compare the forces, independent of operating speed, fluid density and impeller dimensions, they are represented by a non-dimensional force notation per Stepanoff (8):

$$Kr = \frac{Fr}{D_2 b_2 \Delta p}$$

Figure 3 shows the ranges for Kr defined by the results of all shop testing. In figure –3– a wide variation is found for the unsteady radial force levels. This is not a result of deviations in the force prediction results, but is generated by the enormous variety of impeller/casing hydraulic parameters and geometrical deviations tested. (327 Impellers in 47 multi stage pumps and one single stage process pump, representing 46 different impellers designs and 12 different volute designs.) In these graphs, also results obtained by Kanki (3) and Guelich et al (4) are plotted to compare the new test results with previous data generated by direct measurement techniques. The results of synchronous force level presented in figure 3, are pure hydraulic forces, since mechanical unbalance and probe-runout were substracted.

From the tests it was observed that the dependence of the non dimensional synchronous force level with flow rate is speed-dependent. Force predictions, for variable speed test showed clearly that the non dimensional synchronous force level tends to increase at large flow rates and at high operating speeds. Non uniformities, both in impeller and casing produce the synchronous force components. Due to non-uniform pressure distribution and momentum changes, synchronous forces are generated. Without momentum influence the synchronous force component will have the same character as the Q-H performance curve. At moderate flows and speeds the extra forces generated by non-uniform momentum change will give almost constant synchronous forces. At large flows and speeds, the contribution of non-uniform momentum forces results in larger synchronous forces at larger flows and speeds. Figure 4 shows the contribution of non-uniform pressure and momentum on the total force level. Figure 4 shows also non dimensional forces predicted for a large boiler feed pump, tested over a large speed range above 4000 rpm. Results are indicating the influence of non-uniform momentum on the non-dimensional force notation.

All other force components (i.e. subsynchronous and vane passing forces) are independent of operating speed, when they are made non-dimensional per Stepanoff (8). A non-dimensional force notation for the synchronous force level, independent of operating speed, must have an additional fluid momentum term.

From all the force predictions it became evident that only the vane passing forces have a quite constant circular character. This means that in an orthogonal vertical-horizontal representation, both directions have the same amplitude. The synchronous and subsynchronous forces have a more elipsoidal character. This depends for the subsynchronous forces on the flow rate of the pump. At part load conditions much larger forces are generated towards the volute tongues, while at BEP flow conditions the force levels are almost identical. This indicates that low frequency forces are generated by stall and recirculation forces, but also due to the volute casing by part load flow separation at the volute lips. The synchronous forces have also an elipsoidal character, and the force amplitude ratio remains constant over the entire speed and flow range. Whether horizontal or vertical forces are larger depends for the synchronous forces only on the casing hydraulic deviations. Figure 5 shows results obtained for a large 4 stage boilerfeed pump, with impellers of 420 mm, tested with a variable speed motor in the shop from 4000 to 6000 RPM. From these results the circular force behaviour of the vane passing forces and elipsoidal force character of the subsynchronous (1/10-1/5xspeed) and synchronous forces are clearly observed.

Finally the synchronous hydraulic force levels differ tremendously with mechanical synchronous forces. For the 47 pumps tested, the mechanical unbalance Kr values ranged from .001 up to .009, while the hydraulic unbalance Kr values ranged from .02 up to .20.

5 UNSTEADY HYDRAULIC IMPELLER FORCES VERSUS
   HYDRAULIC PARAMETERS

Unsteady impeller forces are also related to hydraulic parameters of pumps. The research project determines these unsteady hydraulic forces focussed on following parameters :

* Impeller specific speed.
* Impeller suction specific speed.
* Discharge angle vanes.
* Number of impeller vanes.
* Impeller casing lip clearance.
* Impeller shroud casing clearance (maroti effect).
* Cavitation.
* Pumping versus turbine mode (reverse running pumps).

Results were obtained both from tests with the 47 double case pumps during performance testing as well as from the special tests with the single stage horizontal overhung process pump as described in reference 5.
A few test results are discussed below.
There is a strong influence of discharge angle on the unsteady impeller forces at vane passing frequency. Figure 6a shows the relation obtained after a least square curve fit of all test results, at several flow rates. It appears that increase in discharge angle, decreases the vane passing force component.
There is a large difference between the vane passing force components in a pump versus a reverse running pump (hydraulic turbine) at BEP and part load conditions. Figure 6b shows a comparison obtained from a 5 stage double case pump with 260 mm impellers running both as pump and as hydraulic turbine during shop performance testing. It is quite clear from the comparison that the turbine produces much larger vane passing forces. Finally, figure 6c shows an overview of the influence of the number of impeller vanes, on the unsteady hydraulic impeller forces at vane passing frequency. The unsteady impeller forces at vane passing frequency reduce with an increased number of vanes, both at BEP and part load conditions.

6 UNSTEADY HYDRAULIC IMPELLER FORCES VERSUS
   HYDRAULIC DEVIATIONS

Deviations in hydraulic geometry of casing and impeller, affects directly the synchronous force components. During testing of the 47 double case pumps, all hydraulic dimensions were carefully measured. This allows to correlate hydraulic deviations with unsteady hydraulic forces. Furthermore, the single stage process pump used as special test pump was subject to several hydraulic deviations using metal polyester. Figure 7 shows some typical deviations introduced. By all this testing several correlations of hydraulic deviations versus unsteady impeller forces were made :

* Volute casing symmetry.
* Volute area deviation.
* Volute angle deviation.
* Impeller vane location.
* Impeller inlet areas.
* Impeller inlet angles.
* Impeller discharge angle.

From all variations imposed on the hydraulic lay-out of the single stage process pump it appeared that only some deviations have important effect on the unsteady hydraulic forces.

Large influence was observed from :

- Impeller vane location.
- Impeller inlet geometry.
- Volute casing symmetry.

All other deviations also influence the unsteady hydraulic forces at the impeller, but to a lesser extent as the above mentioned deviations.

Figure 8a shows some typical results obtained for the influence of vane spacing on the synchronous force component. Also indicated is the influence of the volute casing symmetry on the synchronous force level, figure 8b.
It appears that the elipsoidal force character of the synchronous impeller forces is generated by casing and impeller deviations. The difference between two orthogonal synchronous force levels is generated by casing deviations. The remaining pure circular force level is generated by the impeller deviations.
Deviations in impeller inlet geometry influence both subsynchronous and synchronous force levels. Figure 8c shows an overview. Non-uniform blade loading due to inlet deviations of the impeller have a large effect on synchronous force levels. Also stall and recirculation forces, at part load conditions are amplified by inlet deviations, resulting in larger subsynchronous forces. The results of this research project show that instead of only controlling mechanical residual-unbalance during pump manufacturing also control of hydraulic dimensions will reduce impeller excitation force levels.

7 REDUCTION OF HYDRAULIC IMPELLER FORCES

Increase of pump reliability and pump life are heavily related to pump vibration levels. All results shown that certain force levels increase tremensously at part load and run-out flow conditions. This means that vibration levels will increase at part load and run-out conditions.
Control of the unsteady impeller forces will reduce overall vibration levels, and therefore increase lifetimes. Some practical solutions are given below.

a) Reduced flow kits.

Due to plant operation at lower capacities, pumps originally designed for large flow rates tend to operate at part loads nowadays.
It is evident that this will introduce larger unsteady impeller force levels. Figure -3- shows that especially vane passing and low frequency (subsynchronous) force levels are sharply increasing. This will give an increase in vibration level, decreasing lifetime. Furthermore, pump efficiency will be lower.
Figure -9- shows a Q-H curve of a single stage, single suction process pump, operating at 50% flow reduction and 45% reduction in head. Normally this head reduction is throttled by closing the discharge valve. Efficiency is than lowered from 72% to 55%.
An attractive solution is using a smaller size impeller, operating at best efficiency point for the reduced flow and head duty, together with a special volute insert in the stationary casing to adjust the extra space created. This will increase efficiency, as indicated in figure -9-, up to 68%. Furthermore, the pump

operates again at BEP conditions, giving lower unsteady radial forces, improving vibrations and eventually pump lifetime. Advantage of the "low flow kit" is that the old conditions can easily be obtained by mounting the original impeller and disassembly of the extra volute or diffuser insert. Also the investment of this "low flow kit" is much smaller compared to installation of a complete new, smaller pump size.
Figure -10- shows more details of the "low flow kit" for the overhung, single stage process pump. In this case a quarter volute.

b) Control of vane spacing.

Impeller vane spacing is very important in control of unsteady synchronous force levels. There are at present two ways to cast impellers:

- Patterns for blade geometry.
- Patterns for waterway geometry.

Both ways are used in pump industry, but they have a different effect on blade spacing. Using perfect patterns for the blade geometry can give enormous deviations in blade spacing. Using patterns for the hydraulic waterways, will give correct vane spacing. Blade thickness will then vary, but this can be perfectly balanced on a dynamic balancing machine. It is therefore preferred to have castings using the patterns for the waterway geometry. Precision castings using patterns for the blade geometry can still give large deviations in vane spacing and therefore synchronous force level.

8 CONCLUSIONS

Ranges are defined for subsynchronous, synchronous and vane passing force levels, using results of 47 multistage double case pump performance tests.
Prediction of the forces is done with a cheap indirect modal method. It is concluded that impeller unsteady forces are related to three main factors:

* Pump flow rate.
* Hydraulic parameters.
* Hydraulic waterway deviations.

Two effective ways are discussed to reduce some typical force levels at BEP and part load conditions :

* Low flow kit.
* Impeller vane spacing versus cast process.

These design modifications will lower unsteady force levels, resulting in reduced vibration levels and increased lifetime.

9 ACKNOWLEDGEMENT

The author wishes to express his gratitude to Byron Jackson management for permission to present this paper.

REFERENCES

1) Flack R.D., Allaire P.E., "Lateral forces on pump impellers". A literature review, Shock and Vibration Digest vol. 16, 5-14, 1984.

2) Verhoeven J.J., Gopalakrishnan S., "Rotor dynamic behaviour of centrifugal pumps" Shock and Vibration Digest January 1988, vol. 20, no. 1.

3) Kanki, H., Kawata, J. and Kawatani, T., "Experimental Research on Hydraulic Excitation Force on the Pump Shaft", ASME Paper No. 81-DET-71.

4) Guelich, J., Jud, W. and Hughes, S.F., "Review of Parameters Influencing Hydraulic Forces on Centrifugal Impellers", Paper presented on Seminar "Radial Loads and Axial Thrusts on Centrifugal Pumps", I.mech.E., London (5 Feb. 1986).

5) Verhoeven J.J., "Excitation force identification of rotating machines using operational rotor/stator amplitude data and analytical synthesized transfer function". 11th Biennial ASME design engineering division conference on Vibration and Noise, 27-30 September 1987, Boston, Mass.

6) Desanghere, "Identification of External Forces based on Transfer Function Measurements : Frequency Response Method", Proc. IIIB of the 8th Intl. Seminar on Modal Analysis, K.U. Leuven and L.M.S., (12-16 Sept. 1983).

7) Okubo, N., "The Analysis of Dynamic Behaviour of a Machine under Operating Conditions", Proc. IIIB of the 8th Intl. Seminar on Modal Analysis, K.U. Leuven and L.M.S. (12-16 Sept. 1983).

8) Stepanoff, A.J., "Centrifugal and Axial Flow Pumps - Theory Design and Application, 2nd Edition, Wiley, New York (1957).

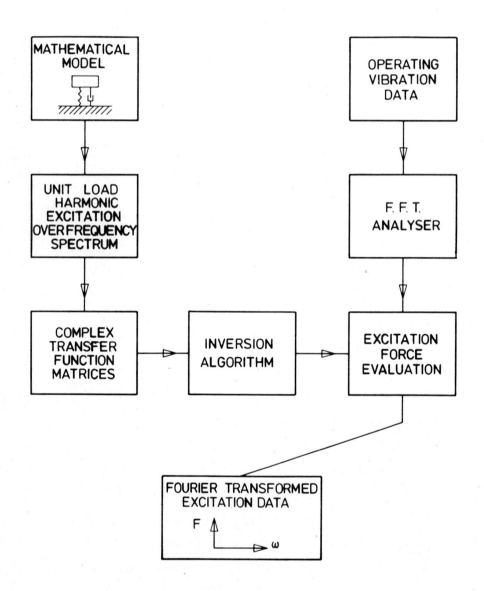

Fig 1   Schematic force prediction algorithm

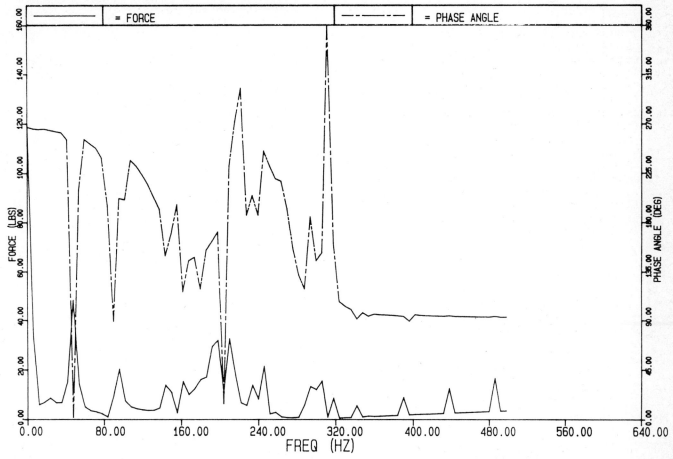

PROJECT DD-863.026.00 6*8*13 L SJA ROTOR/CASING WET PUMP MODEL
FORCE PREDICTION ANALYSIS , TURSEAL CLEARANCE SEAL APPROACH INCL.
ADDDED MASS EFFECTS AND VOLUTE/IMPELLER DATA PER CHAMIEH ET AL

STATION NO. 8   IMPELLER 8-X   AT 2915.00 RPM
PROBE MEASUREMENTS AT SEAL AND COUPLING
PUMP DUTY AT BEP FLOW 660 M3/HR, NO ADDITIONAL MASS UNBALANCE
TRANSLATORY UNIT FORCE EXCITATION

Fig 2   Typical force—frequency spectrogram generated by force
         prediction algorithm

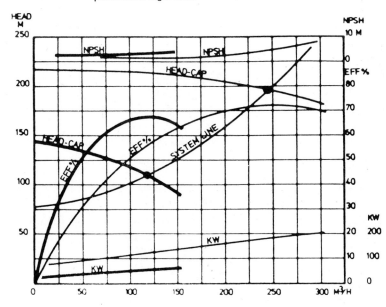

Fig 9   Process pump operating at reduced flow and head conditions
         using a low flow kit

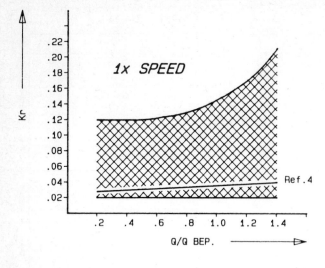

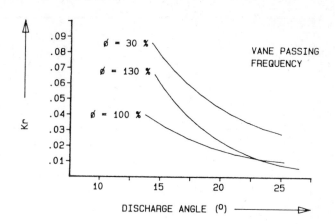

Fig 6a   Influence of impeller discharge angle on force level at vane passing frequency

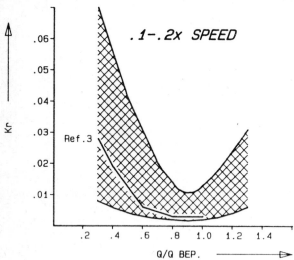

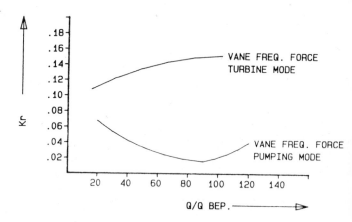

Fig 6b   Influence of pump or turbine mode of operation on force level at vane passing frequency

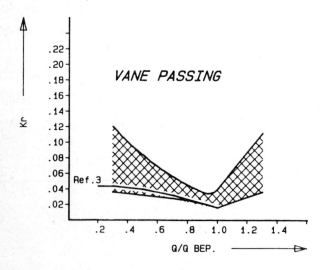

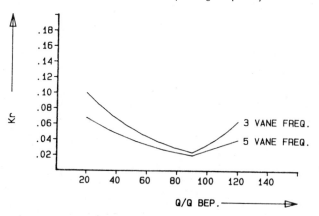

Fig 6c   Influence of number of vanes on force level at vane passing frequency

Fig 3   Ranges of non-dimensionless impeller forces defined by force prediction algorithm

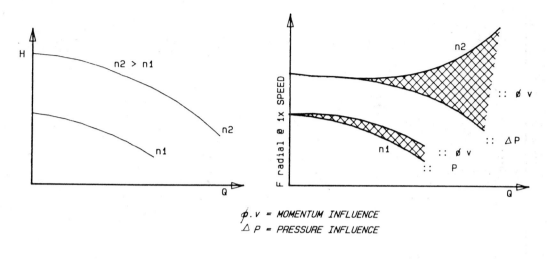

$\phi.v$ = MOMENTUM INFLUENCE
$\triangle P$ = PRESSURE INFLUENCE

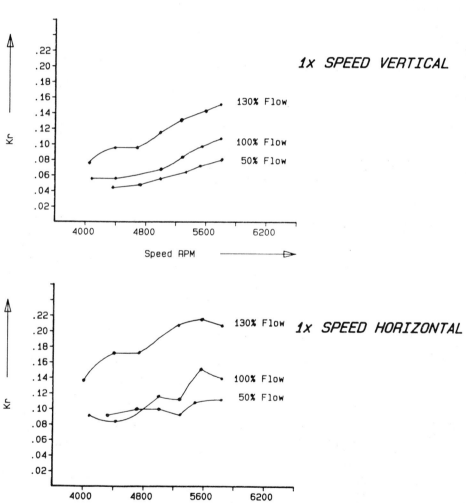

Fig 4   Influence of momenta on synchronous force level, supported by test results

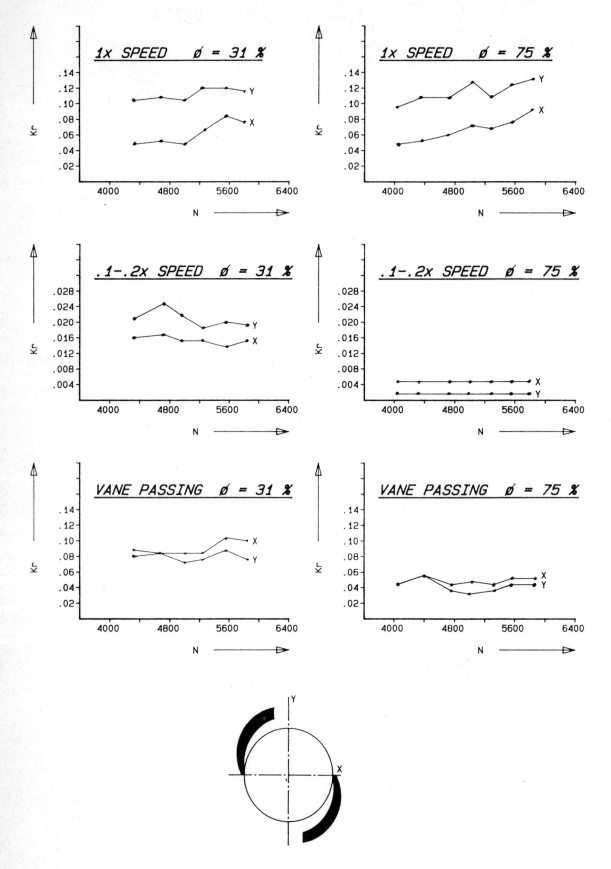

Fig 5   Typical results at several frequencies and flowrates of large four-stage
         boiler feed pump variable speed test; ϕ = flowrate ratio, N = operating
         speed (r/min)

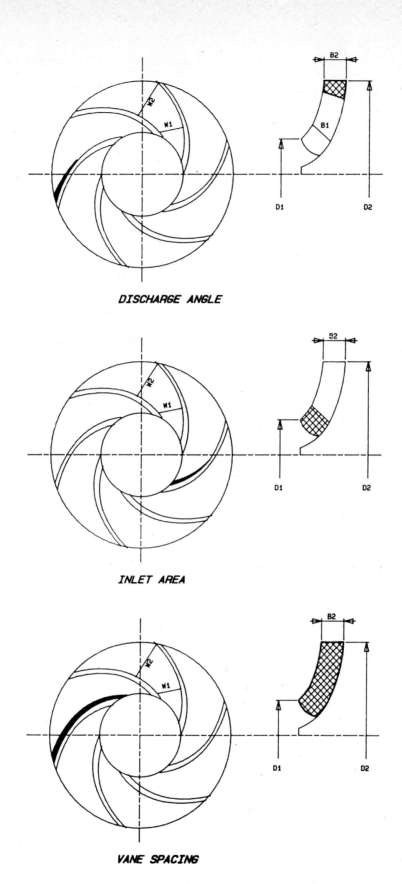

DISCHARGE ANGLE

INLET AREA

VANE SPACING

Fig 7   Typical hydraulic deviations tested on large single-stage process pump

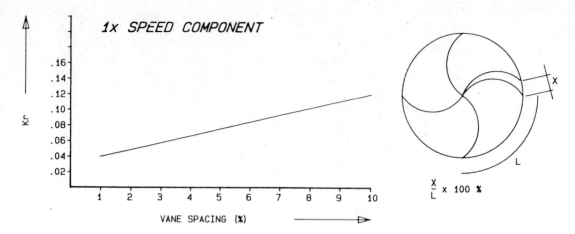

Fig 8a   Influence of impeller vane spacing deviation on synchronous force level

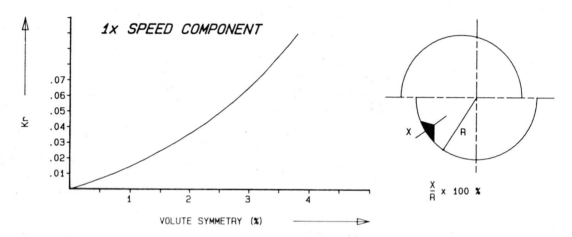

Fig 8b   Influence of volute casing symmetry deviations on synchronous force level

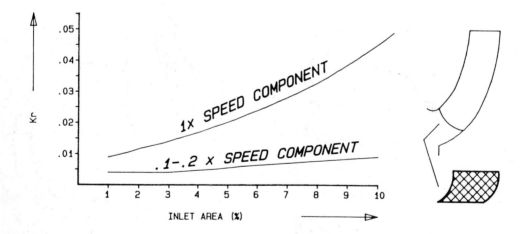

Fig 8c   Influence of impeller inlet area deviation on synchronous and subsynchronous force level

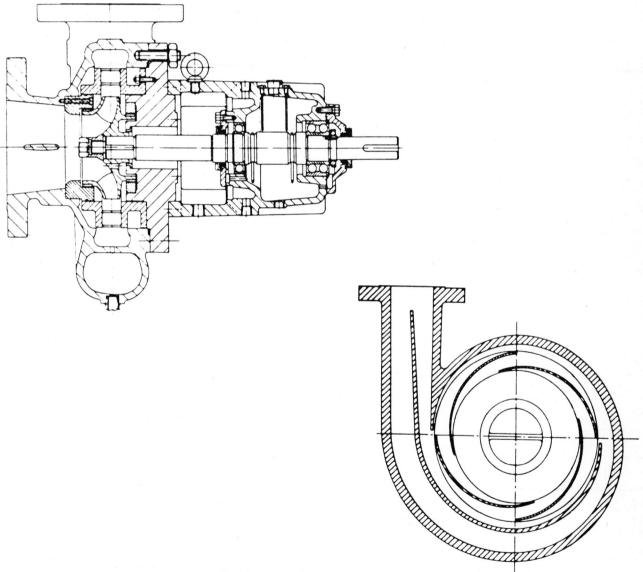

Fig 10   Details of 'low flow kit'

C349/88

Subsynchronous vibration in boiler feed pumps due to stable response to hydraulic forces at part-load

W D MARSCHER, MS, MMEng, MASME, MSTLE, MASTM
Dresser Industries, Harrison, New Jersey, United States of America

SYNOPSIS   Field test results suggest that stable resonance is a more plausible explanation than classical rotordynamic instability in many cases of boiler feed pump subsynchronous vibration. A new model is proposed based on coincidence of one of the lower shaft natural frequencies with the frequency of unsteady hydraulic forces which occur at part load capacities associated with the onset of internal suction recirculation. It is shown how these forces can be present at the stage inlet in the form of a rotating non-axisymmetric pressure field. The rotation of this pressure field occurs at a frequency somewhat below the shaft running speed because of mixing which occurs when the nonrotating inlet flow interacts with rotating impeller internal recirculation eddies which extend upstream from the impeller suction. Using this model, examples are given concerning how the part load flow range, speed range, and severity level at which subsynchronous vibrations occur can be predicted and modified, based on how closely the following coincide on the pump head-capacity map : 1) the system resistance curve, 2) the capacity range of maximum internal suction recirculation forces, and 3) the rotor's first bending natural frequency.

INTRODUCTION

a) Background

Vibration in centrifugal pumps occurs over a range of frequencies, but is usually dominated by the so-called "synchronous" component of vibration i.e. that vibration which occurs with a frequency equal to the pump running speed. Under part load conditions, however, relatively high speed boiler feed pumps have been observed to exhibit strong vibration at frequencies below running speed. Such "subsynchronous" vibration has been observed to originate as shaft oscillation relative to the pump housing, as measured by eddy current proximity probes. It is often associated with field reliability problems such as abnormally high rates of wear in bearings and on annular sealing surfaces, such as break-down bushings, thrust balancing devices, and wearing rings.

In gas-handling turbomachinery, subsynchronous vibration is common over a narrow band of frequencies between 42 and 48 percent of running speed, and can be very destructive. In such cases the vibration is understood to be the result of an unstable feed-back between vibration displacement within various clearance gaps (especially in journal bearings) and some of the fluid forces which occur or increase as a result of this displacement. This type of subsynchronous vibration is known as " rotordynamic instability" (Kirk, Ref 1.) The term "rotordynamic instability" means that after an inevitable perturbation, even an initially perfectly centered, non-vibrating rotor will excite itself at a rotor system natural frequency by responding to vibration-coupled feedback forces.

Solution of rotordynamic instability problems in turbomachines generally focuses on either increasing system damping or minimisation of those forces which can be destabilising because they exhibit "cross-coupling" in a direction which opposes damping (i.e. they create, or occur due to, displacements which lag the cross-coupling force by 90 degrees.) Pursuit of either of these approaches characteristically is time-consuming and costly. The damping or cross-coupling mechanisms which are dominant for the problem vibration mode are seldom certain, and once they are determined their modification often involves substantial design changes and thermodynamic performance compromises.

A number of investigators have associated the rotordynamic instability phenomenon described above with subsynchronous vibrations observed in boiler feed pumps (Childs Ref 2, Pace & Bolleter Ref 3, Makay Ref 4.) Others have proposed it as a probable explanation among several competing possibilities (Massey Ref 5.) Similarities which have been cited between rotordynamic instability and subsynchronous vibration in boiler feed pumps include the following:

1) a dominant frequency below running speed,
2) a relatively sudden beginning or "onset" at a given flow and speed,
3) an onset frequency equal to a shaft bending natural frequency, and
4) high wear rates in bearing and sealing components.

However, in much of the literature and in the cases reported here, there are some important differences in boiler feed pump subsynchronous vibration relative to rotordynamic instability. These differences include:

1) a different characteristic frequency (roughly 85% rather than 45% of running speed),
2) a broad-band rather than narrow-band frequency peak,
3) a characteristic frequency which does not "lock on" to a shaft natural frequency but rather remains closely proportional to running speed,
4) the appearance of frequency peaks called "sidebands" on each side of the running speed peak and its integer multiples or "harmonics", and
5) the ability of boiler feed pumps to run through the condition at which subsynchronous vibrations occur to a higher flow or speed at which the problem disappears.

It is unlikely that rotordynamic instability, at least in its classical form, is responsible for subsynchronous vibrations which exhibit the latter five characteristics. This can be demonstrated by considering the relationship between the various forces on a pump rotor and the rotor vibration which occurs in response to them.

b) Active vs. Reactive Rotor Forces

Forces affecting rotordynamics in pumps may be divided into two categories : active excitation forces and reactive forces (Marscher, Ref 6.) The active excitation forces occur independently of rotor vibration, and therefore are not influenced by its level or frequency. Examples of active forces are internal flow recirculation loads (Fraser, Ref 7), and radial hydraulic " vane pass " loads due to interaction between the rotating impeller vanes and the stationary vanes of the diffuser or volute as discussed by Agostinelli (Ref 8), Hergt and Kreiger (Ref 9), and Makay (Ref 4 .) Since by definition of an active force there is no significant feedback to it from the vibration, active forces cannot excite rotor instability.

By contrast, reactive forces occur as a result of rotor steady and vibratory displacement. An example of a reactive force is the rotor support force provided by the bearings and, in pumps, by the Lomakin Effect in the annular seals (Black, Ref 10.) Examples of reactive forces which do not directly oppose rotor vibration displacement are damping, cross-coupling, and negative direct stiffness such as has been observed in highly eccentric impellers (Brennan, Ref 11.)

c) The Relationship of Reactive Forces to Rotordynamic Instability

Reactive forces which do not act in direct opposition to the vibration displacement provide the opportunity for unstable feedback. This feedback will lead to rotordynamic instability if two conditions are met :

1) the reactive force must act to further increase the displacement which causes it,

2) the cyclic average of the reactive force must exceed the cyclic average of the modal damping force of the vibration mode excited by the reactive force.

Once unstable self-excitation begins at a rotor natural frequency through vibration/ reactive force feedback, it dwarfs any response to active forces, or to reactive forces at non-resonant frequencies. Hence the vibration frequency appears to to "lock-on" to the self-excited natural frequency, regardless of the running speed.

After initiation, the unstable vibration quickly grows to high levels at which rubbing usually takes place within the bearings and annular clearances. Unlike passing through stable resonance between an active force and a natural frequency, running an unstable machine at higher speeds will not bring a return to normal vibration levels unless the rotor support stiffness or damping changes dramatically.

The reactive force most widely recognised as the feedback source responsible for rotordynamic instability is the cross-coupling caused by the hydrodynamic "wedge" which forms in response to "fluid whirl" in eccentric close clearances between rotating and stationary parts. Except for a brief entrance region, the absolute velocity profile across such clearances is roughly linear, equal to the rotor velocity next to the rotor, and equal to zero next to the stationary side of the gap, as shown by Daily and Nece, Ref 12. This results in an average "whirl" velocity equal to slightly less than half of the rotor speed, and is the physical reason why rotordynamic instabilities are nearly always observed at frequencies between 42 and 48 percent of running speed.

As proven both analytically and experimentally by Black (Ref 10), minimisation of cross-coupling is important in BFP rotors running above twice their first bending natural frequency. Black showed that at such speeds the cross-coupling forces, rotating with the clearance fluid whirl at about half the rotor speed, if large enough will overcome damping forces, and rotordynamic

instability will then result. However, BFP rotors flexible enough to allow such a circumstance are unlikely in BFP's designed using modern criteria which focus on stiff rotors, as pointed out by France, Ref 13, and discussed in detail by Marenco (Ref 14). Therefore, rotordynamic instability is not expected in pump designs which have a first bending natural frequency well above half their maximum running speed. If such pumps exhibit subsynchronous vibration problems which do not exhibit most of the characteristics of instability, and in particular occur at a frequency proportional to running speed, then investigators should expect some form of stable forced vibration. The data presented below will be compared to this hypothesis vs. the increasingly common hypothesis that all subsynchronous vibrations in turbomachinery represent rotordynamic instability.

TEST PROCEDURES

a) Methodology

To determine whether resonance, instability, or some other factor is the cause of vibration problems in a boiler feed pump, the following questions should be answered by testing over a range of operating conditions which span the pump service range over which it is experiencing problems :

1. What are the natural frequencies of the rotating and stationary parts of the pump and what is the modal damping associated with them?
2. What are the excitation force frequencies in the actual installation, and do they coincide with the any of the assembled system natural frequencies ?
3. Are the excitation forces within normally acceptable limits ?

In the data presented below, these questions were answered through observation of the vibration vs. frequency response "signature" at various critical locations (in particular between the shaft and bearing housing) combined with the use of Experimental Modal Analysis (EMA).

EMA is a method of vibration testing in which a known force (constant at frequencies within the test range) is put into a pump, and the pump's vibration response exclusively due to this force is observed and analyzed. The actual natural frequencies of the combined casing, piping, and supporting structure in the examples discussed below were obtained from the EMA vibration vs. frequency response using Fast Fourier Transform (FFT) equipment and procedures discussed by Ewins, Ref 15. In addition, the specialised EMA data collection technique of Marscher (Ref 16) was used to determine the rotor natural frequencies at pump operating conditions spanning the range of interest.

As discussed in Ref 15, the EMA response was normalised by the EMA artificial force level at each frequency. The result is a good approximation of the pump's mechanical "transfer function", which quantitatively relates vibration level to excitation force at various locations. Division of the pump's naturally occuring vibration signature by the transfer function at each forcing frequency determined major excitation force frequencies and levels.

b) Format

In the following, vibration test data is plotted in four different forms:

1) "signature" : cartesian plots of vibration amplitude vs. frequency
2) "Nyquist" : polar plots of vibration amplitude vs. excitation/response phase angle (the "lag angle" between the exciting force and the resulting vibratory motion).
3) "time domain" : cartesian plots of vibration amplitude vs. time, similar to a typical oscilliscope trace.

4) "orbit" : polar plots of vibration vs.
time in a plane perpendicular to the shaft
axis.

The frequency and time scales are linear for all
plots. Frequency and time scales are zero to
some maximum value, listed as "X" on the left-hand
side of signature and time domain plots.
The amplitude full scale value on
signature plots is listed as "Y" on the left-hand
side, and is plotted as linear (LIN) or Db
(i.e. base 10 logarithmic.)   In either case the
values plotted are rms averages at any given
frequency, and must be multiplied by 1.41 to
convert them to zero-to-peak amplitude values.
The amplitude scale represents zero-to-peak values
for all Nyquist, orbit, and time domain plots.
The full scale values are in mils/lbf (5.7
micrometres/N) for the impact tests, and for all
other tests is in mils (25.4 micrometres.)

TEST RESULTS

a) 15 MW Boiler Feed Barrel Pump

A layout of a barrel pump of the type used in
various utility and refinery boiler feed services
is given in Fig 1.   In the test installation
discussed here, two pumps, mounted in parallel,
consisted of six stages each ,with double volutes
and five vane 400 mm O.D. impellers. Each shaft
was made with an average diameter of approximately
150 mm, and a bearing span of about 2.3 m. Each
pump used plain journal bearings, was driven by a
steam turbine, and was rated for 15 MW at
5500 rpm. The pumps discharged into a common line.

When either pump was run near half load, broadband frequencies appeared at a frequency of
330 cpm and at sum-and-difference frequencies
330 cpm below and 330 cpm above each of the first
several running speed harmonics, as can be seen in
the natural excitation signature of Fig 2.

The broadband vibrations changed amplitude
depending upon the speed and flow rates of the two
pumps.   The broadband vibrations occured only
when the excitation force frequency of 93 percent
of running speed was within 5 percent of the
rotor's first bending natural frequency , as
determined by EMA impact testing (see Nyquist
plots of Fig 3a), and confirmed by the "peak
hold" test cascade plotting of Fig 3b. The
vibrations were strongest in the vicinity of
the operating conditions for which peak suction
recirculation pressure pulsations were predicted
to occur, per Ref 7. At capacities well above or
below the predicted recirculation pulsations, the
broadband vibrations disappeared.

Fig 4 shows the shaft unfiltered time plot and
orbit during the broadband vibrations. When the
broadband vibrations occured, the orbit was larger
and more unsteady than normal. The time plot
indicates that the unsteadiness was a regular
pulsation, in the form of beating which occured at
the running speed/broadband separation frequency
of 330 cpm.

b) 9000 rpm Four Stage Boiler Feed Pump

Fig 5 shows a four stage BFP which is rated at
4 MW at 9000 rpm. Like the previous pump, this
pump operates in parallel with a pump of identical
design. Under operating conditions, the first
bending natural frequency of this pump is
6300 +/- 500 rpm, depending on the operating point
and the clearances in the annular seals. The pump
is thirty years old, and was designed prior to
engineering knowledge of suction recirculation
(Ref 7). Consequently, each pump stage was
unknowingly designed with a high recirculation
onset capacity of about 100 percent BEP (best
efficiency point) running capacity.

As in the earlier example, the broadband
frequencies occured in the form of a modulating
frequency and running speed harmonic sidebands.
In this case, the main sideband was subsynchronous
at about 80 percent of running speed. The
subsynchronous frequency occurred with a rapid

onset near the first bending natural frequency of
the rotor system, and vibration levels reached the
annular clearances of bearings and annular seals.
However, the subsynchronous frequency did not lock
onto the natural frequency, but instead closely
tracked a constant percentage of running speed.

Fig 6 shows how the broadband frequencies,
including the subsynchronous frequency of 80
percent running speed, occured over speed range
which resulted in an excitation frequency range
centered about the shaft's first bending natural
frequency. Fig 7 shows how the subsynchronous
frequency occurence and severity depends upon
the pump load, peaking at intermediate loads and
disappearing at higher loads approaching BEP.
Fig 8 displays the problem speed vs. flow "window"
on a head/capacity map. This window was observed
to increase in extent and in vibration severity as
annular sealing clearances open up due to wear
over a period of time.

DISCUSSION

a) Interpretation of BFP Subsynchronous Vibrations

The BFP pump vibration problems discussed above
resemble forced vibration, not a rotordynamic
stability problem. A rotordynamic stability
problem involves self-excited vibration which
locks in on the excited natural frequency,
independent of the running speed. There was no
evidence of this in the subject pumps, and to the
contrary the frequency of the subsynchronous
response was a constant percentage of running
speed. Further evidence of forced response was
that the subsynchronous vibration decreased, and
eventually disappeared, at speeds above and below
the speed of maximum vibration. Also recall that
it decreased, and eventually disappeared, at
capacities above and below the capacity for which
maximum suction recirculation forces are
predicted, using the technique of Fraser (Ref 7.)

Concerning the curious occurence of sidebands,
Smith, Ref 17, discusses how in a gearbox the
toothmeshing frequency can have sidebands
spaced by modulating frequencies equal to the
speed of the input or output shaft. Misalignment
causes the amplitude of the each tooth meshing
event to be slightly different than the previous
one. The amplitude modulated gearbox spectra
shown in Ref 17 are very similar to the broadband
components of the subject pumps near the onset of
recirculation. However, the observed modulation
frequency not of running speed, but of 7 through
20 percent of the running speed indicates that the
modulating force in the pumps must slip relative
to the shaft suggesting that the source is fluid
dynamic. This suggestion is reinforced by the
broadband nature of the sidebands, since fluid
dynamic phenomena tend to be broadband due to
turbulence, while mechanical phenomena should
lead only to narrowband integer harmonics of the
forcing frequencies.

An explanation that fits all aspects of the data
is that the vibration was caused by resonance of
the first bending natural frequency of the shaft
with an asymmetrical pressure field induced by
suction recirculation.   Considerable progress has
been made in recent years in the understanding and
prediction of flow recirculation phenomena at off-
peak capacities, such as Fraser (Ref 7), and
Schiavello (Ref 18.)   At certain speeds and
flows, internal flow recirculation phenomena have
been observed to create pressure and velocity
fields at the impeller inlet which rotate
subsynchronously at the frequencies associated
with impeller rotating stall, at 70 to 90 percent
of running speed (Soundranayagan, Ref 19, and
Kammer, Ref 20.)

It is possible that one of the mechanisms for the
asymmetry in the pressure field is rotating stall
in the impeller, as has been suggested by other
recent investigators, such as France, Ref 13, and
for compressors by Wachel, Ref 21. This is
supported by the strongest sideband occuring one
modulation band below the running speed, which can
be interpreted as indicating that the problem

pressure field rotates in the direction of the impeller, with a slip factor relative to the impeller of 7 to 20 percent. The direction and amount of slip are characteristic of rotating stall in the suction of an impeller, per Ref 19 and 20.

Generally, impeller stall and other pressure fields associated with suction recirculation would be expected to be mild, because they occur in the suction, where pressures are low and the area for them to act on is relatively small. Therefore, suction recirculation with or without impeller stall normally would not be expected to cause severe vibration. However, at the operational condition of peak rotor vibration, the recirculation pressure field rotational frequency in each case was within ten percent of the shaft's first bending natural frequency, inducing resonance.

This suggests a methodology for predicting such subsynchronous vibrations will occur, and if they occur how severe they are likely to be. If the pump is run back enough on the system discharge resistance curve of head vs. capacity such that it passes through the capacity range at which suction internal recirculation forces are expected to be maximum (normally between 50 and 75 percent of best efficiency flow, per Ref 7), at least some subsynchronous vibration of the type described in this paper   becomes probable, with a frequency which slips roughly 15 percent relative to running speed . Further, if the rotor first' bending natural frequency is close to 15 percent below the speed that the pump must run at when it encounters such high suction internal recirculation duty points, then the subsynchronous vibration is likely to be severe, and possibly damaging.

Since the above examples were existing installations, the least expensive, least risky solution to avoid the damaging multiple coincidence of suction recirculation, operating speed, and shaft natural frequency was to reprogram the external recirculation and discharge controls to avoid long-term operation of the pumps at any combination of speed and flow which would result in recirculation flows in resonance with the shaft first critical speed.

An alternative solution is to use a device developed at the author's company called a stop vane ring. This consists of an impeller suction-side stationary wearing ring, extended inward radially and then slotted axially at the I.D. to form a configuation resembling a set of inlet guide vanes, except that the vanes are mounted very close to the leading edges of the impeller vanes. If properly designed the stop vanes do not significantly alter the head/capacity characteristics of the pump, but prevent formation of rotating pressure fields at the impeller inlet. They do this by destroying the front edges of the suction recirculation flow eddies as they grow in size and attempt to extend out the front of the impeller where they could interact with incoming suction flow to cause the problem flow fields.

This solution is appropriate for pump designs during development or for installations in which the operating procedure cannot be altered. It has been used successfully on a number of low vibration, quiet service shipborne boiler feed and condensate pumps.

CONCLUSIONS

1. The part-load subsynchronous vibration problems in boiler feed pumps reported in this paper represent simple resonances rather than rotordynamic instability.

2. The boiler feed pump subsynchronous resonances appear to have been caused by coincidence between the rotor's first bending natural frequency and a rotating, non-axisymmetric pressure field at the impeller inlet. It is believed that this field is the result of interaction between incoming flow and extensive suction recirculation eddies which form near the O.D. of the eye of the impeller at certain speeds and below BEP capacities.

3. The recirculation resonance phenomenon may be responsible for many of the subsynchronous vibration problems experienced by boiler feed pumps of various manufacture. If so, these problems can be solved by relatively simple changes in pump operation, or in the configuration of the inlet of each stage.

REFERENCES

1. KIRK, R.G., Evaluation of Aerodynamic Instability Mechanisms for Centrifugal Compressors, ASME 85-DET-147, c. 1985
2. CHILDS, D., Rotordynamic Instabilities in the SSME Hydrogen Fuel Pump, Proc. Rotordynamic Instability Problems in High Speed Turbomachinery, Texas A&M Univ., 1984
3. PACE, S.E., FLORJANCIC, S., BOLLETER, U., Rotordynamic developments for High Speed Multistage Pumps, Proc 3rd Pump Symposium, Texas A&M Univ, 1986
4. MAKAY, E., SZAMODY, O., Recommended Design Guidelines for Feedwater Pumps in Large Power Generating Units, EPRI Rpt CS-1512, Sept 1980
5. MASSEY, I., Subsynchronous Vibration Problems in High Speed Multistage Centrifugal Pumps, Proc 14th Turbomachinery Symposium, Turbomachinery Laboratories, Texas A&M Univ, 1985
6. MARSCHER, W.D., The Effect of Fluid Forces at Various Operation Conditions on the Vibrations of Vertical Turbine Pumps, Proc. IMechE, Radial Loads and Axial Thrusts on Centrifugal Pumps, Feb 5 1986
7. FRASER, W.H., Centrifugal Pump Hydraulic Performance and Diagnostics, Pump Handbook, McGraw-Hill, c. 1985
8. AGOSTINELLI, A., NOBLES, D., MOCKRIDGE, C.R., An Experimental Investigation of Radial Thrust in Centrifugal Pumps, J. Engrg. for Power, Trans ASME, v.82, 1960
9. HERGT, P., KRIEGER, P., Radial Forces in Centrifugal Pumps with Guide Vanes, Proc IMechE, v.184 Pt 3N, 1970
10. BLACK, H.F., Effects of Fluid-Filled Clearance Spaces on Centifugal Pump Vibrations, 8th Turbomachinery Symposium, Texas A&M Univ., c. 1979
11. JERY, B., ACOSTA, A., BRENNAN, C., CAUGHEY, T., Hydrodynamic Impeller Stiffness, Damping, and Inertia in the Rotordynamics of Centrifugal Flow Pumps, Proc Rotordynamic Instability Problems in High Speed Turbomachinery, Texas A&M Univ, 1984
12. DAILY, J.W., and NECE, R.E., Chamber Dimension Effects on Induced Flow and Frictional Resistance of Enclosed Rotating Disks, ASME Journal of Basic Engineering, March 1960
13. FRANCE, D., Rotordynamics Considerations in the Design of High Speed Centrifugal Pumps, Proc Indian Pump Mfrs Assn, March 1987
14. MARENCO, G., Boiler Feed Pump Wet Critical Speeds Beyond the Pump Operating Range, to be published in Proc. IMechE Conf on Vibrations in Rotating Machinery, Edinburgh, 13-15 Sept 1988
15. EWINS, D.J., Modal Testing : Theory and Practice, Research Studies Press, Wiley NY, c. 1984
16. MARSCHER, W.D., Determination of Pump Rotor Critical Speeds During Operation through Use of Modal Analysis, Proc ASME 1986 WAM Symposium on Troubleshooting Methods and Technology, Anaheim Cal, Dec 1986
17. SMITH, J.D., Gears and their Vibration, Marcel Dekker Pub, c. 1983
18. SCHIAVELLO, B., On the Prediction of the Reverse Flow Onset at the Centrifugal Pump Inlet, Proc ASME Gas Turbine Conf and Fluids Engrg Conf, New Orleans, March 1980
19. SOUNDRANAYAGAN, S., Unsteady Operation of Axial Impellers at Part Load Conditions, Proc. Indian Pump Manufacturers Assoc., March 1987
20. KAMMER, N., RAUTENBERG, M., A Distinction Between Different Types of Stall in a Centrifugal Compressor Stage, ASME Journal Engrg for Gas Turbines and Power, Jan 1986
21. WACHEL, J.C., SMITH, D.R., Nonsynchronous Forced Vibration in Centrifugal Compressors, Turbomachinery International, Jan/Feb 1983

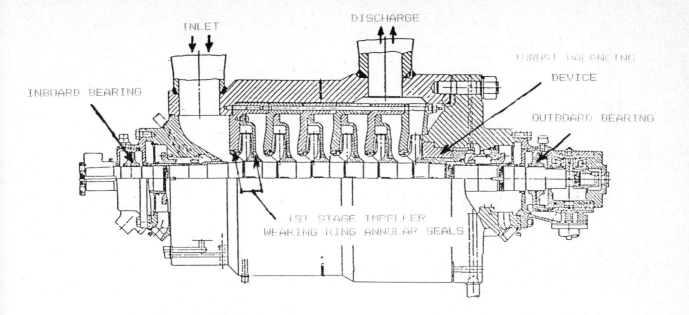

Fig 1   Six-stage centrifugal pump configuration

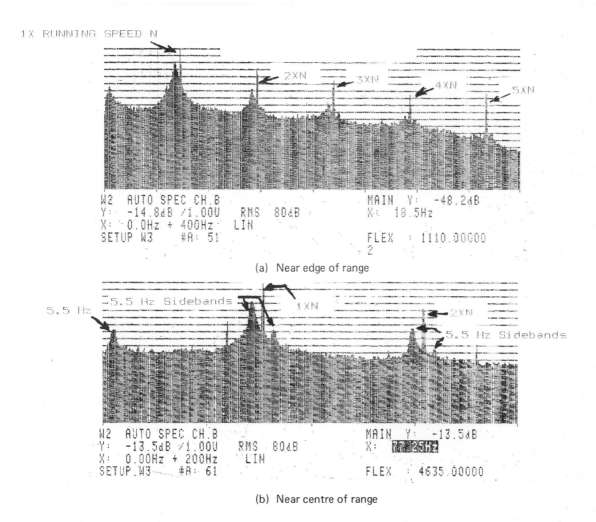

(a) Near edge of range

(b) Near centre of range

Fig 2   Subsynchronous vibrations over limited range of flow and speed

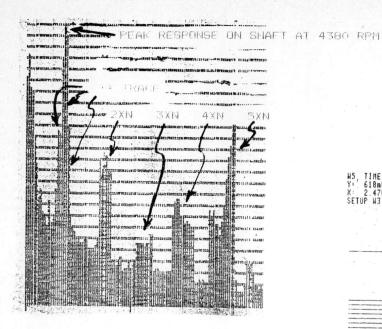

Fig 3b 'Peak hold' plot for 4000–5000 r/min speed transient

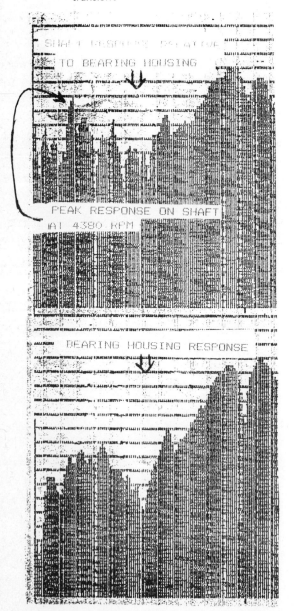

Fig 3a 'Modal' plot of response to artificial excitation

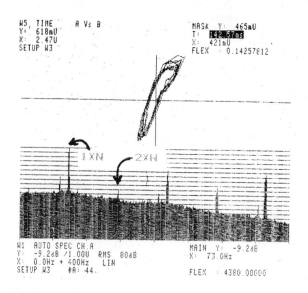

(a) Just prior to subsynchronous vibration

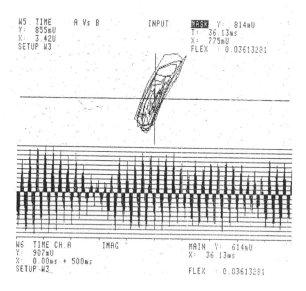

(b) During strong subsynchronous vibration

Fig 4 Orbits and frequency spectra

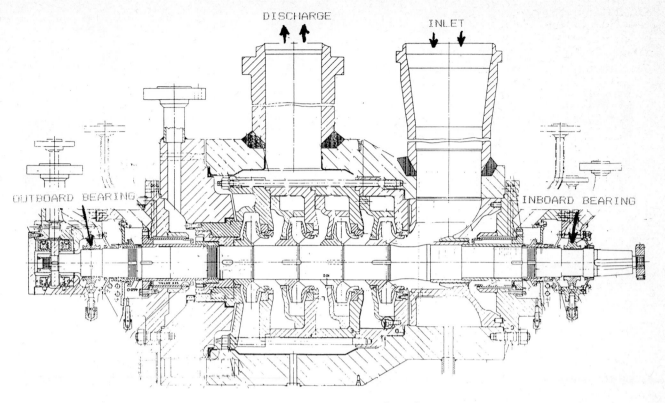

Fig 5 Four-stage pump configuration

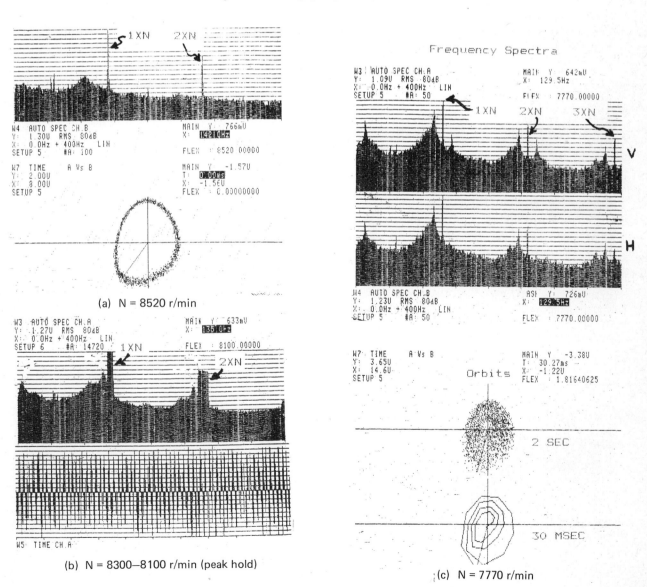

(a) N = 8520 r/min

(b) N = 8300—8100 r/min (peak hold)

(c) N = 7770 r/min

Fig 6 History of subsynchronous vibration versus speed N

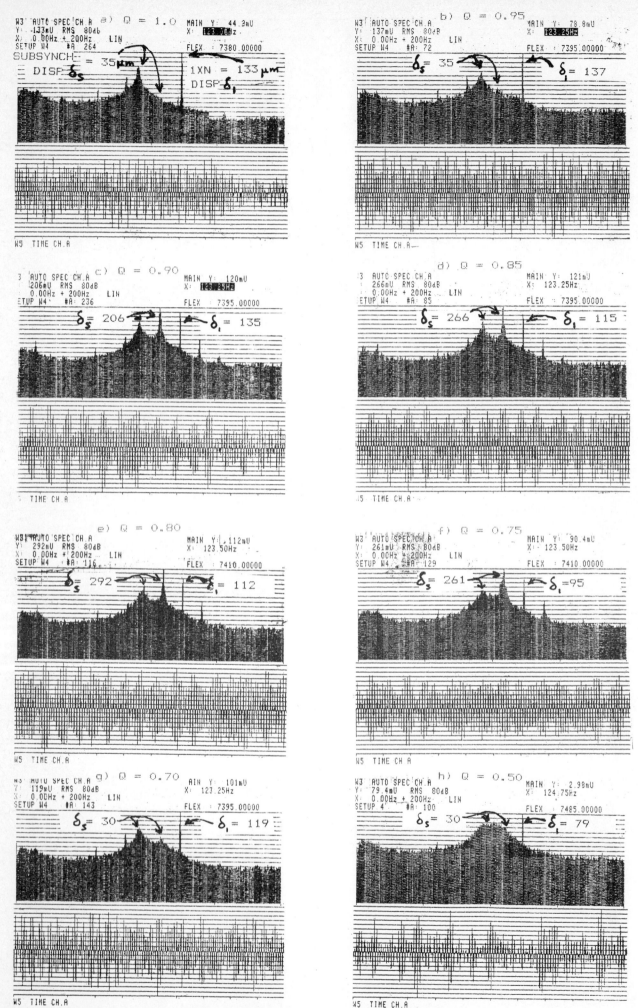

Fig 7   History of subsynchronous vibration versus flow capacity as
         fraction of flow at best efficiency operating point

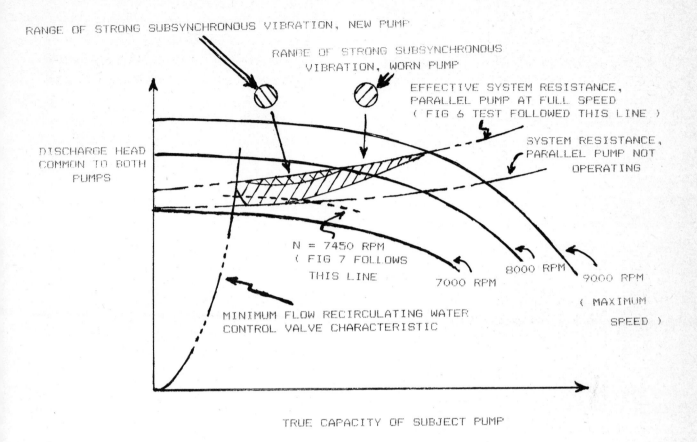

RANGE OF STRONG SUBSYNCHRONOUS VIBRATION, NEW PUMP

RANGE OF STRONG SUBSYNCHRONOUS
VIBRATION, WORN PUMP

EFFECTIVE SYSTEM RESISTANCE,
PARALLEL PUMP AT FULL SPEED
(FIG 6 TEST FOLLOWED THIS LINE)

SYSTEM RESISTANCE,
PARALLEL PUMP NOT
OPERATING

DISCHARGE HEAD
COMMON TO BOTH
PUMPS

N = 7450 RPM
(FIG 7 FOLLOWS
THIS LINE

7000 RPM

8000 RPM

9000 RPM

(MAXIMUM

SPEED)

MINIMUM FLOW RECIRCULATING WATER
CONTROL VALVE CHARACTERISTIC

TRUE CAPACITY OF SUBJECT PUMP

Fig 8   Subsynchronous vibration 'window'

C350/88

The development of a solution to part-load instability in a modern power station feed system

W J ALISON, BEng, CEng, MIMechE and **M C CARTLEDGE**, BSc, CEng, MIMechE
Central Electricity Generating Board, Barnwood, Gloucestershire
D FRANCE, BSc, MSc and **J P WEST**, BSc(Eng)
Weir Pumps Limited, Cathcart, Glasgow

SYNOPSIS: Adverse experience with feed system pipework vibration in a large power station led to extensive investigations. The results of the investigations, consequential impeller development work, mathematical modelling of complex pipework systems and design guidelines for future plant are discussed.

1 INTRODUCTION

Severe feed system pipework vibration has been experienced at a large CEGB coal fired power station over a number of years. Various attempts were made to deal with the problem, including the addition of an enhanced restraint system and culminating in a detailed series of tests aimed at identifying the cause and promoting a satisfactory solution. The first part of this paper covers the investigatory tests which demonstrated a link between pipework vibration and low flow operation of the system, the plant modification that resulted and the proving tests to demonstrate its effectiveness.

This experience is not unique, the phenomenon has occurred in other CEGB/SSEB power stations and worldwide [Refs. 1,2,3,4]. A common factor to all the experience is that the potential problem has not been predicted at the plant design stage. It has not been until the plant was running that the phenomenon has manifested itself. Hence, scope for dealing with it has inevitably been restricted and solutions adopted have in the main been palliatives to ameliorate the adverse effects rather than deal with the cause.

The desirability of preventing a recurrence of the problem on future plant is clear and the paper describes programmes of work undertaken to achieve this objective through :-

(a) Validation of a method and production of a software package for the prediction of the stability of hydraulic oscillations in complex piping networks and assessment of system sensitivity.

(b) Visual cavitation rig tests on high speed feed pump impellers to determine data on the compliance of cavities formed at the pump suction, a necessary input for (a), and also to determine those features of impeller and pump suction geometry which influence the onset and intensity of inlet flow recirculation.

2 HISTORICAL EXPERIENCE

The power station concerned has six 660 MW generating units. It was constructed in two halves, the first unit was commissioned in 1973 and the last in 1986. Throughout most of this period the plant suffered from intermittent severe feed system pipework vibration, the pipework involved being that on the suction and discharge sides of the feed pumps between the deaerator and the feed regulating valves. The arrangement of pipework and feed pumps is shown schematically in Fig.1.

Reports indicated that vibration usually occurred with the unit at part or zero load or during operations involving pump changeover. Investigations of the problem in the early stages, resulted in the addition of pipework restraints and subsequently revised operating procedures were introduced in order to avoid the problem. These measures, particularly the enhanced restraint system, were necessary to enable the plant to be operated within tolerable levels of vibration. Inquiries at other operating stations where identical feed pumps were installed indicated that no similar problems had been experienced there. Vibration problems, however, persisted through to the latter stages of this particular project and therefore, because of concern for the long term integrity of the pipework and the desirability of reducing the maintenance burden resulting from the consequences of vibration (detached lagging, damaged pipework restraints, broken small bore piping), the tests referred to earlier were conducted.

3 INVESTIGATORY TESTS

The tests were designed to establish the conditions which gave rise to vibration and to assess the effects of various parameters on vibration severity as follows:-

(a) For a range of start/standby feed pump speeds the discharge flow to the boiler was varied by throttling at different valves, i.e. feed regulating valves, discharge isolating valves etc., and various combinations of valves both in series and parallel (Fig.1).

(b) With the plant set to a vibrating condition the number of feed suction pumps in operation was varied.

(c) The leak-off systems of the two start-up/standby feed pumps were interconnected with throttling facilities such that up to 50% leak-off flow was available to one feed pump. Leak-off flow was varied for a range of feed pump speeds and forward flows to the boiler.

The conditions of testing were generally as follows:-

(i) Turbine-generator spinning at no load (3000 r.p.m.).
(ii) Boiler drum pressure 105-110 bar.
(iii) Deaerator temperature 105°C.
(iv) Two feed suction pumps in operation (apart from (b) above).
(v) One s/s feed pump operating on leak-off.

Throughout the tests significant parameters were recorded and observations made of plant behaviour.

4 TEST FINDINGS

The major findings from the tests may be summarised as follows:-

(a) Vibration of the pipework occurred at low forward flows to the boiler. It increased in severity as the forward flow increased and then ceased abruptly as flow increased beyond a particular value.

(b) Whilst vibration occurred in both suction and discharge pipework it was most severe in the pipework between the feed suction pumps and pressure stage pumps where it was predominantly axial with a frequency of 2 Hz.

(c) The severity of vibration was not affected by the number of feed suction pumps in service.

(d) Vibration occurred regardless of the method of throttling forward flow, with the same trend as described in (a). However, when the throttling was split between a feed regulating valve and the feed pump discharge isolating valve, discharge pipework vibration was suppressed, but vibration remained on the suction side.

(e) Vibration severity increased with pump speed.

(f) Increasing the feed pump leak-off quantity eliminated the vibration.

Examination of the recorded data revealed that the common factor always associated with vibration was the presence of flow fluctuations at the feed pump inlet. These fluctuations were associated with pump throughput. Increasing pump throughput either by opening a feed

regulating valve or increasing the leak-off quantity eliminated the fluctuations and vibration.

During periods of peak vibration there were associated increases in system pressure fluctuations measured at the feed suction pump discharge manifold and feed regulating valve inlet manifold (Fig.1).

All these observations led to the conclusion that the pipework vibration was due to an acoustic resonance of the hydraulic system being excited by low flow fluctuations at the feed pump inlet.

The increasing intensity of the vibration as flow initially increased could be due to reduced system damping as the feed pump discharge non-return valve, which was located close to the pump discharge, lifted thereby allowing the magnitude of unstable oscillations to increase. The abrupt cessation of vibration observed was due to the feed pump flow increasing beyond the range within which inlet flow fluctuations occurred.

From the test results it was possible to define a boundary of sensitivity in terms of pump speed and flow (Fig.2). At total pump flows below those defined by the boundary, the system would vibrate whilst at greater flows it would remain stable. It can be seen that this boundary ranges from 33% b.e.p. flow at minimum speed to 48% b.e.p. flow at maximum speed.

5 REMEDIAL MEASURES

The test findings indicated three possible routes to remedy the problem:-

1) Develop a more benign feed pump impeller.

2) Alter the dynamic characteristics of the hydraulic network.

3) Increase the feed pump recirculation quantity.

Option 1) was pursued as a longer term strategy and is described in Part 7 below.

Option 2) could be achieved by placing restrictive elements such as the feed regulating valves close to the feed pump discharge, the existing location of these valves being some 40 m downstream of the feed pumps. However, layout considerations and cost implications precluded this as a viable option.

Option 3) therefore was pursued as the preferred means of achieving an effective solution in the short/medium term.

On the basis of the test results it was decided to provide both s/s feed pumps with 50% recirculation capacity if this could be accommodated without impairing operational flexibility. A careful review of operational requirements during unit hot/cold starts, shutdown, pump changeover and emergency start-up led to the conclusion that this could be achieved.

The s/s feed pumps were already provided with 25% leak-off capacity by means of two 12.5% capacity on-off valves and pressure reducing vessels. Various options were considered for provision of the additional 25% including a parallel on-off leak-off system, a parallel leak-off system with modulating control and a recirculation loop around the pump with modulating control.

The latter was chosen as the preferred option and is illustrated in chain dot on Fig.1. With this system up to 25% duty point flow (75 kg/s) could be recirculated from the discharge to the suction side of the pump, the flow quantity being controlled by means of a modulating control valve taking a flow signal derived from an orifice plate located in the pump suction pipework. This orifice plate also provided the flow signal for leak-off valve operation.

The objective of the scheme was that the recirculation loop coupled with existing leak-off facilities should maintain total pump flow outside the low flow region, where vibration was known to occur, under all normal operating conditions.

It was recognised that the introduction of an additional control loop could produce interactions with the existing boiler feed pump speed control/boiler flow control loops. Also, potential destabilising interactions between leak-off valve opening/closing events and recirculation valve action during both single and multi pump operation required investigation.

The investigations were carried out using computer simulation techniques. The studies yielded controller settings and flow settings for leak-off valve and recirculation valve action to secure stable operation.

The desired value of total pump flow was set at 55% duty point flow (165 kg/s). The original leak-off valve settings had been:-

i)   Signal to open on falling flow = 33% duty point (100 kg/s).

ii) Signal to close on rising flow = 65% duty point (195 kg/s).

It was found necessary to reset these to:

i)   Signal to open on falling flow = 45% duty point (135 kg/s).

ii) Signal to close on rising flow = 75% duty point (225 kg/s).

On falling load therefore, the recirculation valve would commence to open when pump flow fell to 165 kg/s and continue to open with decreasing flow to the boiler, until fully open. This would be followed by the leak-off valves opening when total pump flow fell to 135 kg/s. The final condition of the fully de-loaded pump would be with recirculation and leak-off valves fully open and pump throughput according to speed as defined by the 50% leak-off curve on Fig.2.

On rising load the recirculation valve would commence to close as flow exceeded 165 kg/s followed by leak-off valve closure at 225 kg/s.

6   PROVING TESTS

The modification described above was installed on both s/s feed pumps of one unit and a series of proving tests were conducted to demonstrate its efficacy.

The tests all concentrated on low load, low flow conditions and included:-

i)   Repeat conditions from the first test series which had resulted in vibration.

ii)   Floating boiler safety valves. (Known to have given particularly severe vibration).

iii) Unit start-up.

iv)   Unit shut-down.

v)   Pump change-over.

No vibration resulted from any of the above tests and the operability of the plant with up to 50% recirculation flows was demonstrated.

7   IMPELLER INVESTIGATIONS - VISUAL CAVITATION RIG TESTS

The pressure stage feed pumps are high speed (7800 r.p.m.) two stage machines with a high impeller power density. It should be appreciated that the feedpumps were designed some 15-20 years ago. Visual cavitation rig testing has been performed on the suction impeller which revealed the existence of relatively large vapour cavities within the impeller passage over the full duty flow range. Interpretation of these visual cavity curves suggested that suction recirculation was also present over a large proportion of the duty flow range.

The observed frequencies of the unstable oscillations were typical of those normally associated with cavitation surge. Furthermore the necessary ingredients for surge, reverse flow and pre-whirl together with cavitation, were known to exist. However the pump head versus flow characteristic had no regions of positive slope and it was therefore felt that the instability was unlikely to be a classical cavitation surge. Nevertheless there have been reported case studies where reverse flow has been responsible for violent flow unsteadiness. As the suction impeller has a high power density and the recirculating flow draws its energy from the rotating blades and dissipates it entirely in the suction pipe it was felt worthwhile setting up a rig test programme to fully investigate the recirculation and cavitation characteristics of the impeller and to identify and measure the energy in any low frequency disturbances that might be responsible for the pipework vibration.

A first stage s/s feed pump impeller was installed in the cavitation visualisation test rig in Weir Pumps Ltd. Research Laboratory. The test loop comprised a vertical pump in which the impeller eye faced upwards and was viewed through a perspex window in the top cover. The impeller was driven from below at a speed of 1300 r.p.m. by a direct drive variable speed motor.

The closed loop test circuit incorporated a storage tank in which a free surface water level was maintained. The circuit pressure and hence NPSH was controlled by varying the air pressure above the water level in the storage vessel. The water was continually deaerated so that visibility into the impeller eye was not impaired by air coming out of solution. After initial tests a suction stand pipe was incorporated into the test loop to try to induce low frequency surging. The inlet chamber, suction guide and inlet guide vanes replicated the inlet configuration of the feed pump.

The cavity length and volume were assessed under stroboscopic lighting by observation through the perspex window against a graduated scale on the impeller blades. The onset of impeller inlet recirculation was detected by recording the differential pressure between the inlet chamber static pressure tapping and a small bore pitot tube facing upsteam and mounted at the impeller eye outer diameter and was verified visually using wool tufts close to the impeller inlet.

A small bore pitot, large bore pitot and a Kistler pressure transducer, mounted as close as possible to the impeller inlet, were used to monitor pressure pulsations. Cavitation inception was detected acoustically using a Kistler pressure transducer mounted in the perspex cover monitoring high frequency noise levels.

The suction surface cavity length, noise level rise and generated head were measured on test at reducing cavitation number at ten flow coefficients. From this data cavitation coefficients for acoustic inception, cavity lengths from 4 mm to 30 mm and 3% head drop were derived and were as shown on the summary graph (Fig.3a). The site operating conditions are superimposed on the summary graph which confirms that over the normal operating flow-range the pumps operated with very large suction surface cavities. The estimated cavity volume and square root of the differential pressure as determined on test were plotted against flow coefficient for fixed cavitation numbers (σ). Fig.4 shows a typical result. The onset of recirculation occurred where the graph of the square root of the differential pressure vs flow coefficient deviated from a straight line at a flow coefficient almost equal to full load flowrate and was independent of cavitation number.

The visible cavity volume reached a maximum at a flow coefficient of approximately 55% of full load flow coefficient, then reduced rapidly at lower flows reaching very low levels (<100 mm^3) at the original leak-off flow coefficient. The rapid reduction in cavity volume corresponded with a transition from established blade cavitation to intermittent blade cavitation. The very strong recirculating flow at the impeller inlet serving to effectively suppress the cavitation.

Attempts to induce surging by varying the circuit impedance were totally unsuccessful. Furthermore spectral analyses carried out on the pressure transducer signals demonstrated the absence of any dominant frequency at the lower end of the frequency band to generate the conditions observed at site.

A similar programme of tests was carried out on several other high speed feed pump impellers taken from operating CEGB stations with known satisfactory system part load operation. The onset point and intensity of suction recirculation varied but the impeller in question could not be considered untypical in this respect. In no case could significant low frequency oscillations be excited.

To summarise then, the suction impeller is demonstrated to have inlet flow recirculation existing over almost the entire operating flow range. This is particularly intense over the flow range where system instability is observed. However there are no significant low frequency oscillations associated with the recirculating flow and cavitation surge most certainly does not exist on the rig.

The impeller test evidence therefore supported the analysis that a large compliant volume of vapour does exist at impeller inlet due to extensive suction cavitation. This can be characterised as shown in Figs. 5a/5b for use in the system analysis program described in Part 8. Recirculation in itself is not necessarily detrimental but where cavitation exists increasing recirculating flow intensity with reduction in pump throughput can have the effect of improving flow incidence locally, thereby suppressing cavitation and reducing the total vapour volume. This results in a positive sloping volume v flow characteristic (high dV/dQ), that is destabilising cavity compliance coefficients. Whether this will result in unstable oscillations depends on the tuning of the acoustic impedances of the whole system.

A replacement impeller was designed using modern design techniques. The objectives were to reduce the cavity compliance coefficients to insignificant values at all flows above the normal 25% leak-off flow. The design features influencing the onset and intensity of inlet recirculation are manifold but for the required duty the inlet: outlet diameter ratio is dominant. The theory is discussed in Ref.5 but in basic terms above a critical diameter ratio the onset of inlet recirculation is triggered by discharge recirculation and modifications to the inlet geometry will not have a significant influence. From the experimental evidence it was clear that the existence of inlet recirculation is quite acceptable so long as substantial cavitation is avoided. The visual cavitation test curves for the replacement impeller (Fig.3b) demonstrate that significant vapour cavities have been eliminated from the impeller suction over the entire flowrange.

The impeller will now be installed and run at Site to determine the influence on system stability.

8. THEORETICAL STUDIES

The severe vibration described earlier demonstrates only too clearly the complexity of the dynamic response of modern feed systems and the need for a dynamic design method for pipework which considers both the mechanical and fluid system characteristics.

Guidelines on dynamic design methods for pipework, considering the mechanical vibration natural frequencies and optimum use of restraints and damping devices already exist; in themselves they have been found insufficient to totally eliminate vibratory motions of concern. It is clear that the prediction of the fluid oscillations is essential if pipework is to be designed with any confidence.

This section describes the investigation undertaken, using a computer program based on Ref.6, to validate a method for the dynamic design of complex feed system pipework, utilising the findings of the vibration testwork described in Sections 3 and 4.

The method of analysis used for modelling a pipe network, comprising components such as pipe lengths, throttling devices, pump impedances and cavity compliances, is described in Ref.6 and is based on a transfer matrix method. Matrices are calculated for each individual component element and these are then multiplied together in the order specified in the network layout.

The eigenvalue/eigenvector solutions are then derived to establish system natural frequencies, stability data and Pressure and Velocity mode shapes.

A number of simple acoustic test cases served to prove the validity of the method. These test cases were extracted from Ref.7. Two examples are given here, comparing the experimental and theoretical results derived in Ref.7 with the results obtained, in Table 1. The agreement in frequencies is excellent and the mode shape data, not shown, is also in good accord.

Table 1 Acoustic Natural Frequencies of Simple Piping Systems.

| | Example (1) | |
Experimental Results	Calculated In Paper	Calculated From Program
61 Hz.	60.0 Hz.	60.0 Hz.
100 Hz.	98.7 Hz.	98.6 Hz.
146 Hz.	144.6 Hz.	144.5 Hz.
227 Hz.	225.0 Hz.	224.8 Hz.
	Example (2)	
Experimental Results	Calculated In Paper	Calculated From Program
26 Hz.	26.4 Hz.	26.3 Hz.
63 Hz.	62.4 Hz.	62.3 Hz.
101 Hz.	101.1 Hz.	100.7 Hz.

Due to the complexity of the pipework layout in modern power stations many involved interactions can exist. This makes it difficult to assess the significance of the behaviour of individual components on the stability of the fluid oscillations.

A computer model of the feed system was built up from a single pipe to incorporate the feed regulating valve inlet manifold and the leak-off line from one Start/Standby Feed Pump. The final model is shown in Fig.6 and this was believed to be a reasonable representation of the actual power station feed system for which test data was available. Analysis of the results established the frequencies of interest in the different parts of the system. The Eigenvalue solution of the model is given below in Table 2. All the roots here are stable, indicated by the negative critical damping ratio, as there are no de-stabilising influences in the system.

Table 2 Eigenvalue Solution of the Model

Root	Frequency	Damping Ratio
1	0.836 Hz.	− 0.0343
2	0.964 Hz.	− 1,0000
3	2.348 Hz.	− 0.0061
4	2.543 Hz.	− 0.0480

Before proceeding to analyse the feed system in detail, it was considered necessary to investigate the sensitivity of the fluid oscillations to certain component parameters, for example it was important to discover

(i)   whether the position of critical throttling devices such as the feed regulating valve affects system stability,

(ii)   to what extent an unstable pump characteristic destabilises the overall system,

(iii)   the importance of dynamic impedance characteristics of the pumps,

(iv)   the effect of pipework geometry.

The findings from this work are given in the next section.

9 SENSITIVITY STUDIES ON FEED SYSTEM MODEL

(i) Unstable Pump Head - Flow Characteristic

To examine the de-stabilising effect of an adverse pump characteristic, a nominal head drop of 3% was taken as the fall off in pump head at low flow. This had the expected de-stabilising effect, however, it was not of sufficient magnitude to promote system instability. Of the four basic eigenvalues only the one at 2.54 Hz was altered, with its critical damping ratio reducing to −0.0186 from −0.0449.

The calculations were repeated using a head drop of twice the previous value and the system became marginally unstable at low flows. This 6% head drop was felt to be unrealistic for a real pump and although an unstable head-flow characteristic does effect system stability, it would appear not to be a major factor with regard to this system.

(ii) Dynamic Pump Impedance – Cavity in the Pump Suction

The second potential de-stabilising influence would appear to be the presence of a cavity in the feed pump suction pipe or impeller. Some nominal cavity compliance numbers were taken from Ref.(1) for this preliminary investigation. The effects on the system natural frequency at 2.54 Hz, can be seen in Fig.7. Examination of the results shows that the presence of a cavity can have a significant influence on the system stability. It was concluded that detailed investigation into the size of the cavity compliance numbers was needed, if site observations were to be predicted by theory.

10 CAVITY COMPLIANCE/DATA INVESTIGATION

The sensitivity studies performed showed the importance of the cavity compliance number employed. With little or no evidence to support the magnitude of the values used, no attempts at postulating a solution for stabilising the system could be made with any confidence. In order to establish credible values for the cavity compliance numbers for the feed pump first stage impeller, a number of papers that have attempted to quantify cavitational effects were reviewed, Refs. (8)(9)(10)(11). The majority of the methods encountered were semi-empirical relying heavily on test data from purpose built rigs, the data obtained being used to assist with the calculation of the dynamic transfer characteristics of the pump and cavity combined.

For the purpose of this work, a feed pump first stage impeller was inserted into Weir Pumps Ltd. scaled cavitation test rig at Alloa, and visual estimates of the effect of varying pressure and flow were made, as described in Section 7. Graphs were produced for the change of cavity volume with flow and N.P.S.H. (Figs.5a & 5b). From these cavity compliance numbers were calculated for the conditions of interest.

11 THEORETICAL RESULTS AND COMPARISON WITH SITE TEST DATA

Severe vibration was observed at site when the unit was operating with one Start/Standby pump running on a low forward flow, throttling through one Feed Regulating Valve.

Using the cavity compliance data obtained from the test rig, these site conditions were modelled. In their original form the compliance numbers were of insignificant magnitude to promote instability. A scaling factor was applied to the cavity numbers and the resulting eigenvalues emulated the site conditions to a high degree of accuracy (Fig.8).

The effect of splitting the throttling between the feed regulating valve and discharge isolating valve greatly reduced this vibration. Using the cavity compliance data calculated, excellent correlation between site observations and modelled test results was achieved (Fig.8).

The frequency of vibration observed at site was in the area 2.0–2.5 Hz, and this compares well with the modelled frequencies of 2.3–3.0 Hz. However, more importantly, the model predicts with a high degree of accuracy the areas of unstable operation of the system.

With a reliable model having now been validated, the next stage of the study examined the ways in which these unstable zones might be eliminated or at least minimised.

12 ASSESSMENT OF SYSTEM MODIFICATIONS TO ACHIEVE STABILITY

As previously mentioned the major source of damping in the system is the feed regulating valve throttling. The effect of moving the throttling position up the pipeline towards the pump discharge was investigated. The results, Fig.9, show that the closer the application of throttling to the pump the more stable the system became. This postulated the theory that the volume (inertia) of fluid between the pump discharge and the throttling valves influences system stability. As the system is throttled in three places on the discharge side of the pump, the above investigation was repeated with the volume of fluid between the other two valves and the pump discharge line reduced to a minimum. The effect on stability is shown in Fig.9.

13 APPLICATION OF COMPUTER MODELLING TO FUTURE FEED SYSTEM DESIGN

The accuracy of the computer modelling depends heavily on the cavity compliance numbers, and without reliable cavity data, the accuracy of the program is called into question. Thus work is required on a method of estimating cavity compliances for the program to be trusted as a design tool. However with a representative model of a system it can be used to optimise the damping present efficiently and consequently improve system stability.

The empirical data relevant to the impeller investigated could be used for an initial assessment of any future feed system proposals. However, since the design of feed pump impellers is so varied, it would not be sensible to rely on that data alone. Weir Pumps Ltd. have recently reviewed several methods of measurement for pump impeller dynamic characteristics, and it is believed that a method established by N.E.L. (Ref.7) for fluid borne noise silencer evaluation could be used to establish pump dynamic characteristics.

Experience from the work performed on feed system modelling showed that a pump with a positive valued head flow curve will destabilise an overall system, however its effects are secondary compared to those of a cavity. It also showed that the position of the throttling is crucial to the amount of damping in the system, and this leads to the tentative design criterion that the volume (inertia) of fluid between the pump discharge and the throttling positions should be minimised as far as is reasonably possible.

14 IN CONCLUSION

Adverse experience with feed system pipework vibration in a large coal fired power station has led to implementation of a modification designed to avoid low flow operation of the system. The impeller development work which resulted from this experience has been described and also the verification of a method for predicting the stability of hydraulic oscillations in complex piping networks has highlighted its potential value as a design tool to minimise the effects of possible system instabilities by optimisation of plant layout and the positioning of restrictive (damping) elements within the system.

15 ACKNOWLEDGEMENTS

The authors wish to thank Messrs. Weir Pumps Ltd. and the Central Electricity Generating Board for permission to publish this paper.

16 REFERENCES

1. ERSKINE & HENSMAN. Vibration Induced by Pump Instability and Surging. Vibrations and Noise in Pump, Fan and Compressor Installations. I.Mech.E. Sept. 1975.

2. SIMPSON & TRAMSCHEK. Pulsations in Power Station Feed Pump Systems. Conference Publication 4. I.Mech.E. 1974.

3. CHEN, FLORJANCIC and STURCHLER. Reduction of Vibrations in a Centrifugal Pump Hydraulic System.

4. DUSSOURD, J.L. An Investigation of Pulsations in the Boiler Feed System of a Central Power Station. Journal of Basic Engineering. ASME December 1968.

5. FRASER, WARREN H. Flow Recirculation in Centrifugal Pumps. Proceedings of Tenth Turbomachinery Symposium.

6. BLACK, H.F. and LUIS DOS SANTOS. Stability of Oscillations in Boiler FeedPump Pipeline Systems. I.Mech.E., C111/75.

7. SAKAI, T. and SAIKAI, S. Study on Pulsations of Reciprocating Compressor Piping Systems, First Report, Calculation of Natural Frequencies of Complicated Piping Systems. Bulletin J.S.M.E., Vol.16, No.91, January 1973, pp63-68.

8. YAMAMOTO, K. Experimental Study on the Dynamic Behaviour of a Cavitating Centrifugal Pump. I.A.H.R. Symposium 1986, Montreal.

9. YEDIDIAH, S. Certain Effects of Recirculation on Cavitation in Centrifugal Pumps. Proc. Inst. Mech. Engrs, Vol.200, No.A4.

10. DESMET, B. and BARRAND, J.P. Analyse des Fluctuations de Pression a L'aspiration et au Refoulement d'une Pompe Centrifuge. I.A.H.R. Symposium, 1986, Montreal.

11. JEFFERS, D.E. Condensate Pump Surges. C.E.G.B. Scientific Services Dept., NW Region, Service Note.

12. WHITSON, R.J. The Measurement of Acoustic Properties and Transmission Loss of Components in Oil Hydraulic Pipe using a Two Point Technique. N.E.L. Report.

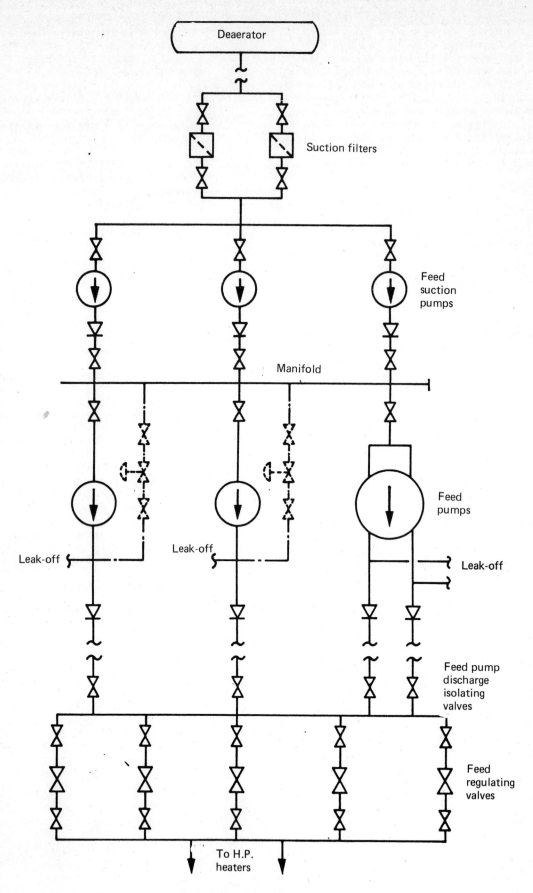

Fig 1   Diagram of feed range pipework

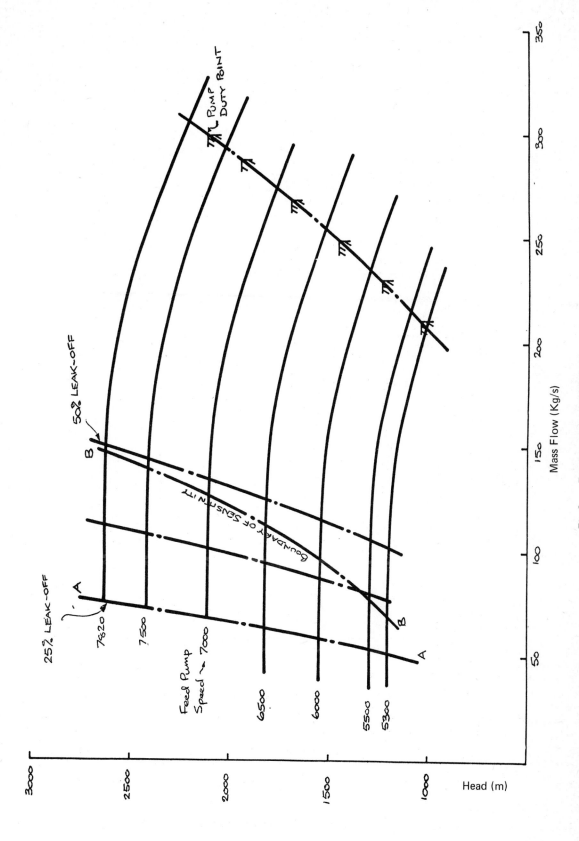

Fig 2   Feed pump characteristics

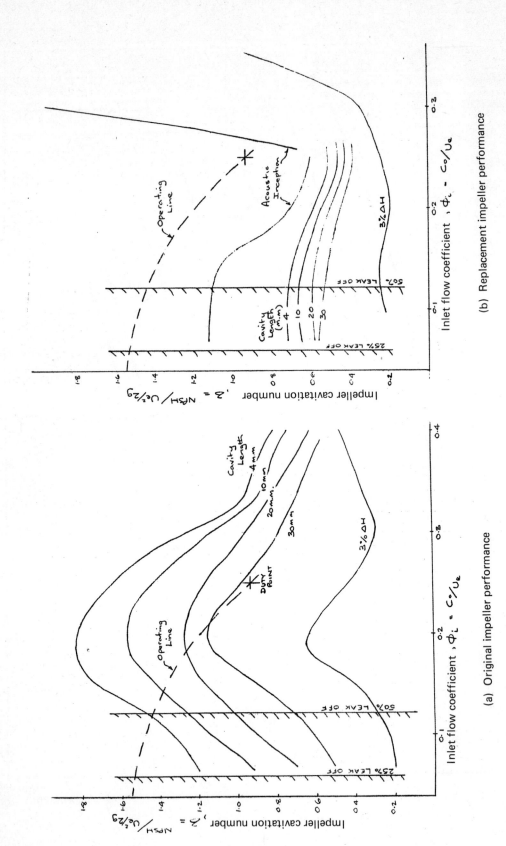

(a) Original impeller performance

(b) Replacement impeller performance

Fig 3   Visual cavitation rig tests

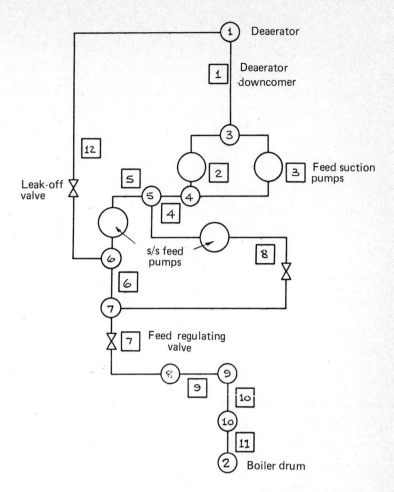

Fig 6   Boiler feed system 'shortened' model

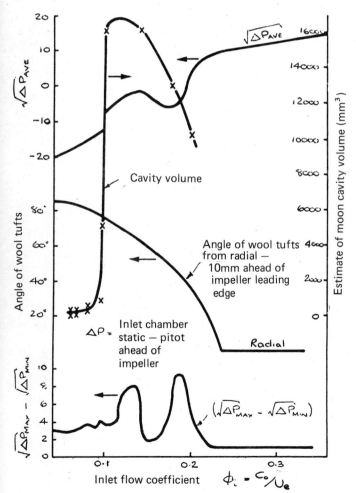

Fig 4   Visual rig performance of original impeller

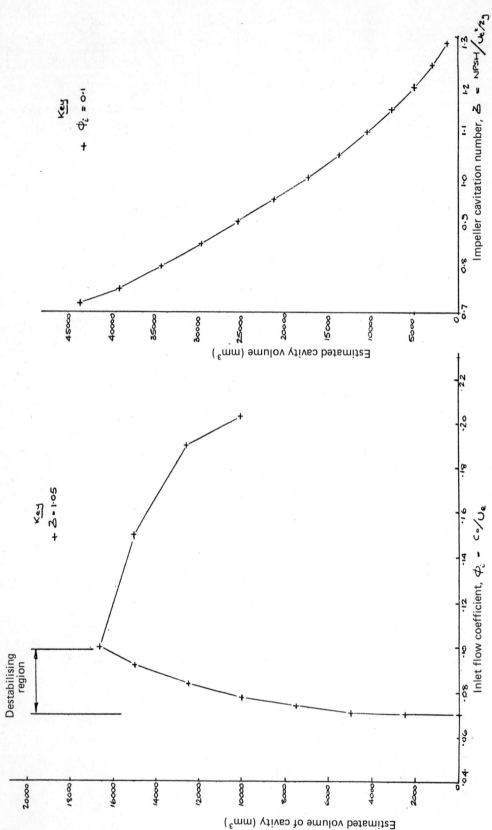

Fig 5   Cavity volumes for original first-stage impeller

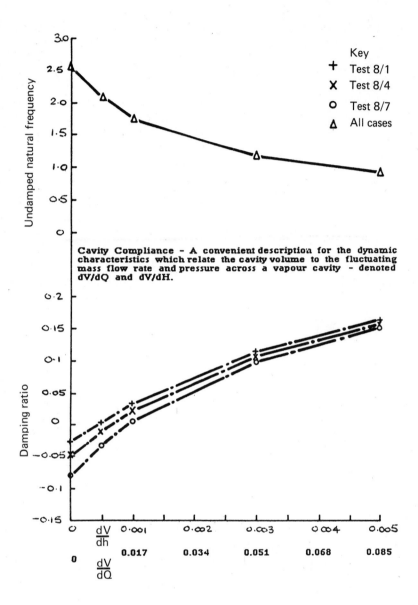

Cavity Compliance – A convenient description for the dynamic characteristics which relate the cavity volume to the fluctuating mass flow rate and pressure across a vapour cavity – denoted dV/dQ and dV/dH.

Fig 7   Cavity compliance sensitivity study

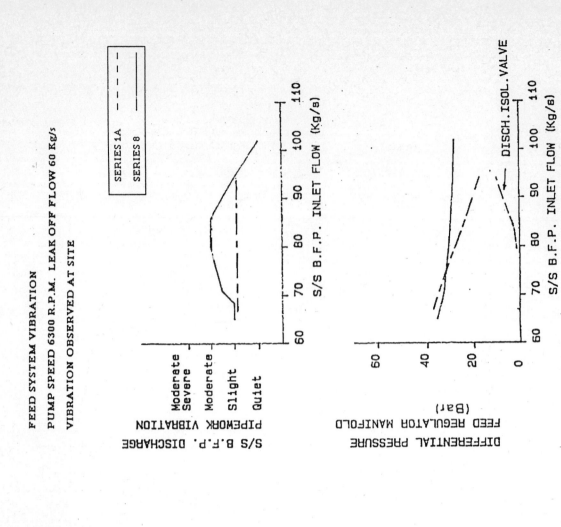

FEED SYSTEM VIBRATION

PUMP SPEED 6300 R.P.M. LEAK OFF FLOW 60 Kg/s

VIBRATION OBSERVED AT SITE

MODELLING OF TEST SERIES 1A AND 8.

Fig 8   Comparison of site test observations with computational modelling results for test series 1A and 8

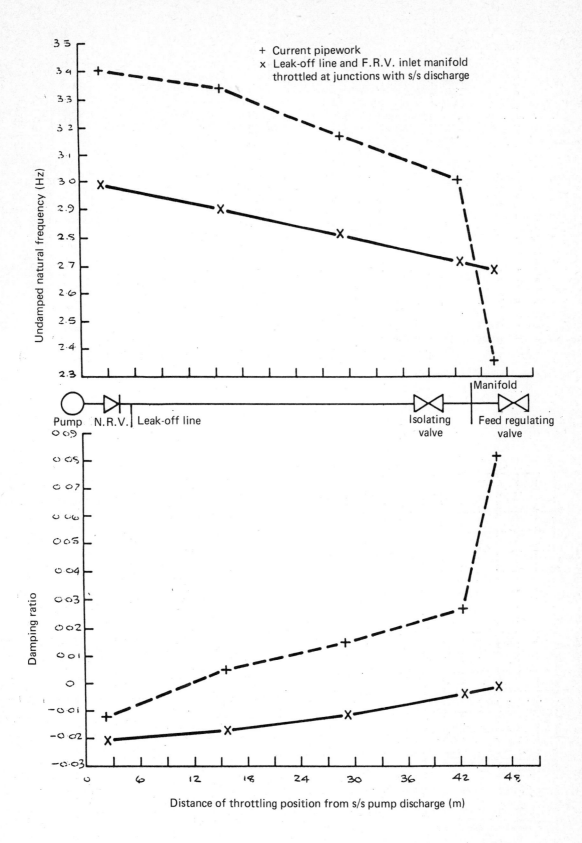

Fig 9   Effect of moving the throttling valve position along the start/standby
         discharge line on system stability for test case 8/4